HOW THE WORLD WORKS

HOW THE WORLD WORKS

A Guide to Science's Greatest Discoveries

BOYCE RENSBERGER

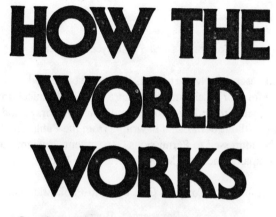

Quill

William Morrow

New York

Library of Congress Cataloging-in-Publication Data

Rensberger, Boyce.
How the world works.
Reprint. Originally published: New York: Morrow, c1986.
1. Science—Popular works. I. Title.
[Q162.R38 1987]   500   86-33236
ISBN 0-688-07293-3 (pbk.)

Printed in the United States of America

2 3 4 5 6 7 8 9 10

BOOK DESIGN BY PATTY LOWY

To Judith, Erik and Joel

CONTENTS

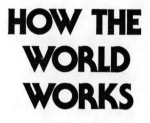

HOW THE WORLD WORKS

INTRODUCTION

This book is for people fortunate enough still to be child-like in their approach to the world around them.

All children are born scientists. They explore their world with relentless curiosity. It's not enough to know what things look like. They must be touched, fondled, banged against other things, even tasted. The toddler who pours the juice onto the floor is making fundamental discoveries about gravity, fluid dynamics, and, for that matter, parental psychology. The third-grader who keeps asking those pesky "why" questions is exploring the link between cause and effect.

Every child goes through these stages of intellectual development, but unfortunately, almost every child also faces repeated discouragement from adults. What mother can tolerate a baby's turning the house into a messy and sometimes costly laboratory? And what father has not ordered his child to stop asking so many questions? Growing up, in part, is the process of having one's natural curiosity beaten down. It is the rare, persevering child who conquers the system and turns amateur interests into a professional career.

Less rare are those people who don't become scientists but who still retain a healthy fascination for things. What makes apples fall? Why is sugar sweet? How do daffodils know when to start sending up their shoots? Why are sun-spots dangerous for astronauts? What would it be like to meet an *Australopithecus*? How could a few molecules acquire the property of life? How did the universe begin? And why?

On behalf of all of us, professional scientists have been asking these questions and thousands of others. They are not unusual in wanting to know how the world works. They differ from the rest of us only in knowing some methods of getting at the answers, methods deliberately designed to minimize the chances of misunderstanding. Scientists have almost always been eager to share the answers they find.

How to Use This Book

This book, then, has two purposes—to bring the childlike reader "up to speed" in the major areas of scientific knowledge from the past, and to serve as a reference book for those who are following the leading areas of current research. This book is *not* a comprehensive encyclopedia of scientific information. There are already many of these on the shelves. Nor is this book much concerned with technology, or the practical application of scientific knowledge. Again, plenty of other books do this quite well. Instead, this book offers a highly selective look at science from two points of view, one to serve each of the book's purposes.

The first of these two viewpoints is a synopsis of twenty-four of the major areas of scientific discovery that shape our modern perception of the natural world, from atoms to human beings. Although specialists have been consulted in the preparation of nearly every entry in this book, and on the selection of the twenty-four major areas of discovery, the final choices were mine. Also mine are the ways I have chosen to explain the selected topics. The level of detail and complexity, which varies among the entries, represents my judgment, after nearly twenty years of presenting scientific information to the general reader, as to how much most people are prepared to wade through in connection with a given topic.

Some of the areas chosen may seem strangely trivial, such as cell theory, the idea that living organisms are made of discrete cells. Actually, the discovery of cell theory was not only an intellectual triumph; it was a concept that made it possible to understand how a complex organism with many specialized organs could grow from one tiny, fertilized ovum that looks like nothing much at all. Other areas are more obviously consequential—the genetic code, evolution, atomic structure, and the ultimate area of research, the origin of all existence in a so-called Big Bang of the Universe.

Each of the twenty-four areas of discovery is described very briefly in Part 2 of the book. From this, it is possible to get a quick sense of the scope of science as a whole. Each description is followed by references to entries in Part 3, where longer articles provide the historical context, factual background, and details of each discovery. If the discovery was predominantly the work of one researcher, it may be presented in the context of a biography of that person. If many people contributed, references in Part 2 will direct the reader to articles on the subject and relevant biographies.

The second purpose of this book is as a reference volume dealing with scientific terms that crop up frequently in popular articles about current scientific research. Part 3 contains explanations of these terms, along with other entries that expand on the twenty-four areas described in Part 2. Articles in newspapers and magazines are often written on the assumption that everybody knows what DNA is or what a supernova is. Quarks and quasars may be presented with the merest of phrases to explain them. Part 3 explains many of these terms. This part also lists seemingly routine terms such as *antibody, hominid,* or *weightlessness,* which are commonly misunderstood. Some entries include diagrams and charts to help explain the underlying concepts. If you want a refresher on the proper place of kingdoms, phyla, and families, check the entry *classification.* If you can't remember which came first, the Carboniferous or the Permian, look under *geologic ages*.

Part 3 also contains brief biographical sketches of nearly all of the most important scientists throughout history, those whose contributions have stood the test of time. This latter consideration has meant, of course, that most of the scientists working in recent years are not included or are presented only very briefly. The chief reason for this is that the stories of their lives, apart from their discoveries, have not yet begun to resonate as memorable themes in the history of science. What's more, there are quite simply too many of them for a book of this nature. There are more scientists working today than there have been throughout all previous eras of history combined. This could mean that many years from now the scientific pantheon will mainly contain scientists from the 1980s. More likely, however, it will mean a weakening of what has been called the Great Man syndrome. Seldom any more are major advances in science attributable to one person. Teamwork has been the rule for some years now. Still, the biographical entries that are included—although some may fall prey to the Great Man syndrome—should give most readers a reasonable familiarity with the history of science.

Part 1 deals with the nature of science itself. What is it that sets scientists apart from anybody else who asks questions about the world? What is the so-called scientific method? As Part 1 describes, it is simply the addition of some methods that can extend childlike curiosity into a creative force that will last a lifetime.

PART 1

What Is Science?

There must be a thousand answers to the question: What is science?, none of them fully satisfying. Some are correct but broad: "Science is the search for knowledge." Some are realistic but not helpful: "Science is what scientists do." Some are idealistic but misleading: "Science is the search for truth through the logical sequence of observation, formation of hypotheses, testing by experiment, and so forth and so on to devise general principles about the workings of the natural world."

All of these definitions make sense as far as they go. The problem is that none of them conveys the highly diversified range of activities and styles of approach of which science is capable. There are scientists who observe and record things—a field biologist watching animal behavior or an astronomer photographing galaxies. There are scientists who conduct experiments—a chemist combining substances in a test tube or a physicist smashing protons together. And there are scientists who just sit and think—a cosmologist calculating the forces operating in the first millisecond of the Big Bang or a molecular biologist trying to figure out how one molecule can tuck itself into the folds of another.

One way to view science is as an intellectual approach, a style of thinking and a set of methods all designed with one goal—to exclude or at least minimize the chance of being misled by an observation. This is really what the scientific method is all about. It is a set of rules for not getting fooled.

Consider this example: You wonder what the chances are that any given birth will produce a boy or a girl. It is a perfectly scientific question, as is just about any question concerning how the world works. There are several ways to find the answer. You could simply observe the outcomes in your own family. You might record that your family has two boys and one girl and conclude that nature is built so that twice as many boys are born as girls. You might. But the scientific method teaches that you run a great risk of being misled if you do not make *many* observations and examine them as a group. It is bad science to state a general rule on the basis of only one observation. There is a high probability that you could get fooled.

So you look at another family and find that the three children in it are all girls. You check a third family and find one boy and one girl. Gradually, as you widen your sample and total up the numbers, you find that the ratio of boys to girls begins to approach 50–50. A strange thing starts to happen: once your group of samples reaches a certain number of families, the ratio never again varies very far from 50–50. At this point the scientific method permits you to stop counting and begin accepting the measured average as pretty close to the truth. (Had the real ratio been 75–25, for example, the ratio in your total sample would have approached those numbers and become stabilized close to them.) The field of statistics is, in fact, a servant of science because it offers mathematical rules for determining ahead of time such things as how many families you have to count to have a high probability that your measured average is close to what it would be if you counted all families.

Now you have produced a little piece of scientific information: The chances of any birth producing a girl are 50 percent, and the same goes for producing a boy. Having made observations that you can accept as reliable, the scientific method now allows you to take the next step—to ask *why* the sexes should be so evenly matched. It is always true in science that one good answer provokes at least one new question.

The usual practice in science when confronted with a "why" question is to think up a "because" answer. You can search your memory for something you already know to be true and which now looks like a good explanation. Or you can make up something. If you do this, you call your idea a hypothesis. "The sexes are equally matched," you might hypothesize, "because everybody should have somebody to love and a 50–50 ratio makes this arithmetically possible."

One of the rules of the scientific method is that the only good hypothesis is one that can be tested. Otherwise you don't know whether it's right or wrong. So scientists try to state their hypotheses with some kind of test, or experiment, in mind. The hypothesis in this example is not very testable. It's the kind of hypothesis that was very popular among the ancient Greeks, however. Objects fell downward, Aristotle taught, because the ground was their rightful place.

The Greeks were pretty smart, but they did not insist on testability as the *sine qua non* of a good idea. Science does. A testable hypothesis for our birth-ratio research project might be as follows: "The sexes are equally matched because there are only two combinations of sex-determining chromosomes that a person can inherit, and the odds that an embryo gets one or the other are 50–50." This can be tested by going out and looking at other species which have the same chromosome situation and observing whether they also produce half males and half females.

Now let us suppose it is a year or two later and that we have had a wonderful time on our grant-supported field trips checking up on the sex ratios of lions and hummingbirds and termites and tuna. And we have gone to the library to look up the papers of other scientists who have already reported sex ratios for other species. Something is wrong. The sex ratios are not all 50–50. Many are, but others are all over the place. Our hypothesis needs work. It isn't as simple as we thought. The scientific method is sending us a message: Go back to the drawing board. Don't count on winning the Nobel Prize this year.

In the meantime, one of our graduate students has been re-examining the human birth ratio and has come up with a flaw that suggests our original conclusions were too simplistic. What we counted originally was simply the number of male and female children from each set of parents. The student now points out that if we group the numbers according to the ages of the children, our 50–50 ratio doesn't hold. The ratio among newborns shows that significantly more boys are born than girls. The ratio moves closer to 50–50 as we move up through older age groups. Adolescents and young adults are close to 50–50. Among older adult siblings the proportion of females has moved ahead of males, and in old age the number of women is very much larger than the number of men. Our graduate student—clearly a promising candidate—has even gotten death-rate figures. They fit the pattern: At every age males are more likely to die than females.

A deeper analysis of the data has revealed a phenomenon that is far more profound than it seemed at the outset. It seems obvious that somehow, nature has arranged things so that the number of males who die before the age of parenthood is offset by a larger number of males born in the first place. Our first hypothesis—the "somebody to love" theory—may not have been so bad after all, because the new analysis supports the idea that the system is built to produce a 50–50 sex ratio during the reproductive years. It obviously did not evolve so that older women would have an equal number of older men in their lives.

Now we are faced with a much tougher research problem, figuring out how the human reproductive apparatus produces more males than females to begin with. This is a genuine scientific mystery. The data given in the example

above are true. In fact, researchers have established that
the preponderance of males is even greater at the moment
of conception than at birth. More male embryos spontaneously abort than do female embryos, and the
imbalance at conception must compensate for this differential mortality as well as the greater male death rate
following birth.

The preceding example has been long, but it has illustrated many elements of the scientific method—the need
for reliable observation of what is happening in the real
world, the formulation of hypothetical explanations (hypotheses) that can be tested, and the performance of a
well-designed test to see whether the hypothesis stands
up. Scientists prefer hypotheses cast in a form that leads
to predictions. Ideally, the hypothesis should suggest an
experiment that has never before been done, and predict
the outcome of this experiment. If the first test of a hypothesis looks good, it still does not ordinarily constitute
proof that the hypothesis is true. Scientists are always
afraid that the test may somehow have concealed some
unanticipated phenomenon that really accounted for the
result. Experiments are elaborately designed to exclude
extraneous factors that could influence the outcome, but
there is always that chance that one has slipped by. As a
result, important findings are not often accepted until
other scientists have repeated the experiment in their
own labs and completed their own search for extraneous
influences. Once the result of the experiment has been
confirmed, scientists begin to believe the hypothesis.

At this point it is necessary to reveal a little inside information about how scientists work, something the textbooks don't usually tell you. The fact is that scientists are
not really as objective and dispassionate in their work as
they would like you to think. Most scientists first get their
ideas about how the world works not through rigorously
logical processes but through hunches and wild guesses.
As individuals they often come to believe something to be
true long before they assemble the hard evidence that will
convince somebody else that it is. Motivated by faith in his

own ideas and a desire for acceptance by his peers, a scientist will labor for years knowing in his heart that his theory is correct but devising experiment after experiment whose results he hopes will support his position. The scientists whose work is described in this book are among the minority who eventually persuaded others that their hunches were correct. Unmentioned are the many whose hunches never bore fruit.

What makes a scientist leap to a hunch that will sustain him through years of inconclusive experimentation? There are many factors, but the most common is the quality often described as beauty or elegance. Certain ideas about how the world works just *seem* right because they are simple and clean. A powerful tradition has grown up in science that favors the simple explanation over a complicated explanation of the same thing. For example, consider the ancient theory that planets and stars moved about the Earth in their different paths because they are carried on concentric, transparent spheres that rotate in different directions and at different speeds. The rules by which one sphere moved did not necessarily govern another sphere. In this view the universe is a complex and cumbersome machine. When Newton proposed his simple laws of universal gravitation and motion, all the seemingly crazy comings and goings of celestial objects could be accounted for by the same few equations. In Newton's hunch the moon, planets, sun, and stars all obeyed one set of rules. The simplicity of Newton's system was appealing. Newton's laws had an elegance, a beauty that did as much to win their acceptance as did any experimental confirmation of their truth. One of the unwritten articles of faith in science is that elegance in a theory is testimony to the theory's truth. Sometimes elegance is misleading, but more often it proves a reliable guide to truth. This has prompted many scientists to wonder why the profoundest truths that govern the universe are perceived as beautiful by the human mind—a product of those very truths at work. Are the laws that science discovers beautiful because our brains are the product of those laws? That is to say, is the human mind the universe's way of achieving self-satisfaction? Or, to look at the situation from the opposite point of view: Does the human mind divide the natural world into beautiful but artificial categories, called laws or theories, simply because that is the only way we can make sense of reality?

Modern particle physics, with its whimsically named charmed quarks and black holes, comes very close to suggesting that conventional human languages are quite inadequate for dealing with how the world works at its most fundamental levels. Take light, for example. According to physics it is both a particle—a little lump of matter—and not a particle. It is a wave, a disturbance in some nonlight form of matter. Every experimental test of whether light behaves like other particles shows that it does. And every experimental test of whether it behaves like a wave shows that it does. Yet to say that light is both a particle and a wave violates the commonsense rules of linguistic logic. But it is true in the real world. What this reveals is that conventional language is simply inadequate for describing at least some aspects of how the world works. This is why so much science, especially fundamental physics, seems odd or counterintuitive unless one speaks mathematics, which many physicists hold is the universal language. Since so few people outside science speak this language, this book struggles to explain things entirely in English.

The degree of rigor common to any specialized field of science varies considerably, usually as a function of that field's history and its ability to do experiments. Scientific specialties tend to go through an early, developmental period, and during their infancy their practitioners can do little more than observe certain chosen phenomena. As the field develops, it accumulates enough observations to permit astute thinkers to begin to perceive regular patterns in the phenomena that are observed—patterns that often betray a hidden force, mechanism, or process. Scientists then begin to formulate hypotheses or, to put it more plainly, try to imagine what might be going on behind the scenes. Having done that, they try to think of experiments that will test their hypotheses.

Chemistry and physics are two of the oldest sciences, and perhaps the two most developed. Each can search centuries of observational data for patterns. Each has well-developed methods of testing hypotheses. And each has been at the game long enough to find hypotheses that predict the outcome of an experiment correctly every time. Hypotheses that do this are often called laws or rules. The phenomena dealt with in chemistry and physics usually perform so neatly that they can be described in mathematical terms. Two hydrogen atoms plus one oxygen atom equals one water molecule. Every time.

Biology began its major period of development much later, and not until the twentieth century was it apparent that certain phenomena in living organisms behave with anything like mathematical precision. As a result, much work in the life sciences remains in the early stages of observation, and the search for patterns in the observed phenomena is still young. Those areas of biology that are closely allied to chemistry—biochemistry and molecular biology, for example—are today approaching the degree of mathematical precision common in the older sciences. This is not true of evolutionary studies or even developmental biology, the study of how individual organisms grow. These branches of biology must settle for hypotheses, theories, and laws that can usually be stated only in imprecise terms. There is not yet a mathematical formula into which you can plug the code from a set of genes and learn, after the equal sign, what kind of organism will result.

Still less developed are the social sciences, whose practitioners search for rules and laws governing the behavior of whole organisms, be they people or porcupines. Scientists working in these areas today must content themselves almost entirely with predominantly observational work. They study cultures and individuals, looking for patterns of behavior that stem from common underlying processes. For several reasons, the social sciences have not enjoyed the same kind of success as other disciplines. First of all, these fields are very new, having started only in the late 1800s, several centuries after biology began to take off. And the object of their study, behavior, is vastly more complex than the workings of molecules and cells. Also, social scientists have proven extremely vulnerable to outside pressures that tend to cast the findings and ideas of their disciplines into political and ideological molds. As a result, topics within the social sciences are often neutralized by rival ideological factions. This is not entirely unique to the social sciences, but the other branches of science have produced large data bases that can support theories without recourse to ideology. Presumably, the social sciences will also some day achieve this. But because of their current lack of definitiveness, this book does not deal with the social sciences.

One more point remains in a discussion of what science is: A definition of what it is not. Science is not technology. Although science and technology are intimately related, they are different. Science is the systematic search for new knowledge; technology is the practical application of that knowledge. Science discovers that electricity can be changed into light; technology engineers a practical light bulb. A technologist (usually called an engineer) can tell you how light bulbs are made; a scientist can tell you why they give light.

Although this book is confined almost exclusively to the achievements of science, it can be useful to anyone interested in technology. Indeed, in today's world, increasingly permeated by arcane technologies, a familiarity with the science that underlies these technologies, may be essential to sane and humane living. Some benign technologies may be frightening simply because their underlying principles are unknown. Other, truly dangerous technologies may seem innocuous to all but the scientifically knowledgeable citizen. In either case, society's mastery of its own existence depends increasingly on a familiarity with how the world works.

PART 2

The Major Advances of Science

This section of the book describes those theories and discoveries throughout the history of science that still stand to explain how the world works.

1. The Big Bang and the Expanding Universe

The universe as we know it is not ageless. It was born between 13 and 20 billion years ago. In 1929 Edwin Hubble, an American astronomer, noticed certain peculiarities in the light from stars and galaxies, and realized that the only way to account for his observations was to conclude that these celestial structures were all flying away from each other at great speed. The universe, Hubble discovered, is expanding exactly as if it were exploding from one central point. The explosion came to be called the Big Bang, and by reckoning backward from the present scale of the universe, astronomers have been able to calculate when it started. In the beginning, all of what we know today as matter, energy, space, and time were contained, in some poorly understood primordial form, within a tiny speck. That speck exploded (nobody knows why), and under conditions of unimaginable temperature and pressure, the Big Bang brought forth everything from atoms to galaxies.

The resultant cosmos is organized into a variety of structures. Planets such as the Earth are among the tiniest of these. Larger forms include stars of several kinds, most of which are organized into galaxies. The universe also contains bizarre objects such as black holes, pulsars, and—perhaps most puzzling—quasars whose energy output seems to draw on a source millions of times more powerful than any that physicists can imagine.

See BIG BANG; HUBBLE; STAR; GALAXY; SOLAR SYSTEM, ORIGIN OF; PULSAR; BLACK HOLE; QUASAR.

2. The Copernican Revolution: The Heliocentric Solar System

The Earth, alas, is not at the center of the universe, nor is it even at the center of the solar system. We, like the other planets, orbit the sun. Though today it may seem a trivial discovery, this finding had a powerful impact on philosophies and theologies that proclaimed human beings to be the pinnacle or centerpiece of creation. Nicolaus Copernicus, a sixteenth-century Polish astronomer, proposed that the Earth circled the sun. A century later, Galileo Galelei, an Italian astronomer, used one of the first telescopes to find the evidence supporting Copernicus' proposal. Johannes Kepler, a German contemporary of Galileo, confirmed the finding and improved it by showing the planetary orbits to be ellipses, not perfect circles.

See COPERNICUS; GALILEO; KEPLER; BRAHE.

3. The Origin of the Earth, Plate Tectonics, and Geological Uniformitarianism

Four and a half billion years ago the Earth was little more than a flying heap of debris left over from the formation of the sun. Gradually, however, the debris organized itself into a dynamic planet. Heat at the center of the Earth melted the debris, allowing most of the heavier elements, such as iron, to sink to the planet's core, while lighter compounds, such as the silicates that make rock, floated to the top and hardened.

The heat-driven processes that shaped the primordial Earth have not stopped. They continue to remodel the planet's surface, pushing chunks of the Earth's crust (or plates) in different directions, with the result that continents move continually, sometimes splitting to open huge rift valleys, sometimes colliding and pushing up mountains, and sometimes rubbing against one another, triggering earthquakes.

Though sometimes sporadic, these processes represent a fundamentally uniform rate of geological activity that is constantly, gradually changing Earth's surface features. This idea, called uniformitarianism, was introduced in the nineteenth century as a radical challenge to the prevailing theory of catastrophism, which held that sudden major events, such as Noah's flood, shaped the Earth's surface.

See SOLAR SYSTEM, ORIGIN OF; PLATE TECTONICS; WEGENER (who introduced the theory); UNIFORMITARIANISM; LYELL (the father of uniformitarianism); CATASTROPHISM.

4. Newton's Law of Universal Gravitation

All bodies in the universe exert a gravitational pull that affects all other bodies. This is the force that binds the moon to the Earth and the Earth to the sun. It is what holds galaxies together. The strength of the pull increases with the mass of the object exerting the pull, but diminishes the greater the distance from the object being pulled. Isaac Newton, a seventeenth-century English mathematician and physicist, worked out formulas to calculate how strong a gravitational pull should be for any given mass or distance.

See NEWTON; GRAVITY.

5. Newton's Three Laws of Motion

These are the three principles of mechanics that govern
virtually all moving objects within the everyday realm of
experience. It is now known that these laws do not hold
inside the atom or for objects moving at velocities near
the speed of light, for which Einstein's relativity theory is
more accurate. Newton didn't do any very extensive experiments in order to arrive at these laws. He simply condensed the accumulated observations of centuries into the
simplest set of statements that could account for everything about moving objects. The laws state (1) that an object in motion or at rest will stay that way unless acted on
by an outside force, (2) that a force is anything that
changes the rate of motion of a body, and (3) that for
every action of a force, there is an equal and opposite reaction.

See NEWTON.

6. Einstein's Theories of Relativity

In 1905 Albert Einstein, a German-Swiss-American theoretical physicist, put forth two theories of relativity—the
special theory and the general theory. A brief summary is
scarcely possible in this space. It appears in the entry on
RELATIVITY in Part 3. Suffice it to say here that the theory of relativity shows that Newton's laws do not hold for
objects that are moving at an appreciable fraction of the
speed of light. When objects go that fast, certain characteristics about them change in the eyes of a stationary observer. Fast-moving objects increase their mass and
shrink in the dimension parallel to their motion. Also,
time appears to slow down; an astronaut who flew about
at nearly the speed of light for a year (as measured on his
watch) could return to Earth and find that several years
had elapsed there. Although these are called theories,
there is no question that they are true. Every experimental test has confirmed Einstein's predictions. Perhaps

the most dramatic test was the atomic bomb, which proved the relativity theory's assertion that matter and energy are the same thing in two different guises, each of which can be converted into the other. Nuclear bombs turn matter into energy. Particle accelerators turn energy into matter.

Perhaps the most counterintuitive of the theory's ideas is that there is really no force called gravity. Instead, Einstein said, space is curved. Just as a golf ball's path across a green is bent by the curvature of the ground, so the curvature of space bends the path of objects moving through space. The curvature of space results from concentration of mass. The Earth, an appreciable concentration of mass, curves the space all around it so that the path of any less massive object moving nearby is bent toward the ground. Gravity, Einstein said, is simply a human concept that makes it easier to understand the visible effects of curved space.

See RELATIVITY; EINSTEIN.

7. Atom Theory

Matter is made of very tiny particles called atoms. Although the Greeks first advanced this idea, it was not until Einstein, writing in 1905, that there was direct proof of the fact. Until then, many people argued that matter was some kind of continuous stuff, and that it wasn't grainy, as something made of small particles should be. Almost a century before Einstein, however, John Dalton, an English chemist, developed the first modern concept of atoms by asserting that each chemical element (iron, carbon, oxygen, etc.) was made of its own kind of atom. Chemistry, Dalton correctly said, was simply the business of binding or unbinding atoms in various combinations.

See ATOM THEORY; ATOM; ATOMIC STRUCTURE; DALTON; EINSTEIN.

8. Atomic Structure: Atoms Are Made of Smaller Particles

Atoms, it turns out, are badly named. The word is Greek for "indivisible," but atoms can be divided. They consist of a nucleus (made of smaller particles called neutrons and protons) surrounded by orbiting electrons. Although electrons do orbit, they do not follow the neat circular paths shown in the old models that make an atom look like a solar system. The electrons stay in no clearly definable path, but instead move within cloudlike shells or regions. Even neutrons and protons, it turns out, are not indivisible. They are made of still smaller particles called quarks, of which there are six basic kinds. The so-called strong force binds the nuclear particles, and the electromagnetic force binds the negatively charged electrons to the oppositely charged nucleus.

See ATOMIC STRUCTURE; RUTHERFORD (who first developed the nuclear model); BOHR (who improved it); QUANTUM THEORY (which governs events in the atom).

9. The Nature of Chemical Bonding

Though the world around us is largely made of atoms, most of the matter we encounter is made of larger objects called molecules. Molecules are combinations of atoms that are bound together in regular and predictable ways. Although millions of kinds of molecules can be formed, there are only two fundamental processes that cause atoms to stick together to form molecules—covalent bonding and ionic bonding. Both involve the electrons that swarm around an atom's nucleus. Atoms that share electrons are bound to one another covalently; atoms that transfer them from one atom to another are bound ionically. The elucidation of the nature of chemical bonding, which grew out of the work of many researchers over at least two centuries, made it possible for scientists to begin to understand the fundamental nature of matter itself.

See CHEMICAL BONDING, THE NATURE OF; ATOM THEORY; ATOMIC STRUCTURE.

10. Quantum Theory and Quantum Mechanics

Along with Einstein's theory of relativity, quantum theory and its development into quantum mechanics is generally considered the greatest of the twentieth century's intellectual achievements. Introduced by Max Planck, the German physicist, and developed by Albert Einstein and Niels Bohr, the Danish physicist, quantum theory says that energy, like matter, comes in particles, or grains. In other words, energy is not a continuously variable thing; when you examine smaller and smaller units of energy, you finally come to the fundamental unit—the quantum—and although quanta come in different sizes, there is no such thing as half a quantum. Despite popular usage of the name to the contrary, a quantum is a very small piece of energy, its effect significant only within atoms. If a quantum of energy (be it light, heat, or any other form) enters an atom, it may be absorbed by an electron, with the result that the electron jumps up to a "higher" orbit around its nucleus. Since this position is unstable, the electron soon drops back to its normal orbit, nearer to the nucleus, and gives up the quantum, or emits it. Quanta are also known as photons, which are the fundamental particles of light. Photons having a certain fixed energy, or energy level (corresponding to an electromagnetic wave of a certain frequency, are perceived by the human eye as light. Since, according to relativity theory, matter and energy are interchangeable, quantum theory explains the behavior of all fundamental entities, whether they are described as quanta of energy or particles of matter.

Quantum mechanics is a development of quantum theory that explains the behavior of particles inside the atom. See QUANTUM THEORY AND QUANTUM MECHANICS; ATOMIC STRUCTURE; LIGHT; (and the scientists who de-

veloped the theory) PLANCK; EINSTEIN; BOHR; HEISEN-
BERG.

11. The Unification of Magnetism and Electricity

Magnetism was known in ancient times as a curious property of a mineral called lodestone. Electricity was
appreciated in the more recent past in the form of lightning, static electricity, and as the product of the battery,
which was invented in 1800. In the 1820s Michael Faraday, an English physicist, discovered that if you move a
magnet near a wire (or vice versa), an electrical current
will flow in the wire. Conversely, if an electrical current
flows through a wire near a piece of iron, the iron will
become a magnet. Magnetism and electricity are two different manifestations of one fundamental force called electromagnetism. James Clerk Maxwell, a Scottish physicist,
worked out the mathematical relationships in Faraday's
discoveries, producing what many consider to be the most
elegant intellectual construction of the nineteenth century
and paving the way for the making of electric motors and
generators. In 1897 J. J. Thomson, an English physicist,
established that electricity consists of a flow of particles
called electrons.

See FARADAY; MAXWELL; THOMSON.

12. The Nature of Light

Isaac Newton, the English physicist, said in the 1600s
that light consisted of tiny particles. In this view, a beam
of light was like a hail of machine-gun bullets. Newton's
contemporary, Christiaan Huygens, a Dutch physicist,
said that light was not particles but waves. In this view
light is more like a ripple on a pond, in which the only
thing that moves away from the source is a disturbance in
the medium—in a water wave, for example, the water
doesn't move horizontally. The wave-particle argument

see-sawed for centuries until Albert Einstein argued early in this century that both were correct. He turned out to be right. Paradoxical as it seems, light has a dual nature, being both a particle (called a photon) and a wave (called electromagnetism)—a statement which concedes that our language is incapable of grasping the truth.

Subsequent discoveries show that light represents only a narrow portion of a much broader spectrum of electromagnetic waves that also includes radio waves, x-rays, and heat.

See ELECTROMAGNETISM; PHOTON; NEWTON; QUANTUM THEORY; EINSTEIN; RELATIVITY (in which the speed of light is the only absolute).

13. The Conservation of Matter and Energy

Matter can neither be created nor destroyed; it can only be changed from one form to another. In its day, that was a bold insight, because even so simple an act as burning a log seemed to destroy matter. But Antoine Lavoisier, a French chemist, would set the record straight. He made very careful measurements in experiments in the late 1700s to show, for example, that the mass of an object that seems to be lost by burning simply goes somewhere else. John Dalton, an English chemist working in the early 1800s, showed that not only was mass not destroyed, but that the original atoms were unchanged; they merely reassembled themselves into new combinations.

The equivalent observation for energy was developed independently by several people. Gottfried Leibniz, a German mathematician, put it forth as a philosophical postulate in the seventeenth century. Hermann von Helmholtz, a German physicist, made it more explicit in the nineteenth century. Everyday examples of such transformations include the conversion of chemical energy (as muscle cells burn food) into kinetic energy (as the muscles lift the arm) and thermal energy (as the muscle heats up). The kinetic energy of the arm, which holds a baseball bat, is transferred to the ball on impact. Some of this kinetic energy becomes heat at the moment of impact (both the ball and bat actually suddenly get warmer), but most of it propels the ball. Friction with the air saps some of the ball's kinetic energy, heating the air in its path. When the ball clears the fence and hits the bleacher seat, all the rest of its energy becomes heat.

Although both principles of conservation are true in the ordinary realm of human experience, Albert Einstein showed in the early 1900s that under special circumstances matter and energy can be converted into one another. Still, however, the total amount of matter and energy in the universe remains the same.

See LAVOISIER; DALTON; HELMHOLTZ; EINSTEIN; RELATIVITY.

14. The Nature of Heat and the Laws of Thermodynamics

Heat is in us and all around us. Even on the coldest day objects still have more than 300 degrees of heat, measuring from absolute zero. Although heat was once thought of as a mysterious fluid, physicists now realize that it is a form of energy that exists in matter as the motion of molecules vibrating and rotating, even as they make up part of the hardest solid. The more the molecules move, the higher the temperature of the matter. Any other form of energy can be converted to heat. Burning, for example, converts the energy stored in the chemical bonds of a fuel, such as wood, into heat. The principle involved is the one discussed in the previous section—the conservation of energy.

One of the most fundamental relationships in classical physics is that of heat (the usual standard for measuring units of energy) and work. Physicists define work as a force moving through a distance. Turn the page and you are using the power of your muscles to exert a force that acts through the distance traveled by the moving page.

Start your car and the energy locked in the molecular structure of gasoline begins to do work. It becomes heat, which expands the gases in the engine's combustion chamber, pushing the piston, which rotates the crankshaft, which, after a series of gears, turns the wheels, and the car moves. Heat (therme is Greek) has been turned into a force that makes the car move through a distance (dynamikos is Greek for power).

The study of the relationship between heat and work has produced the three classic laws of thermodynamics. The first law is simply a formulation of the idea that heat is a form of energy and that, while energy may change form, it is never entirely lost—a state of affairs known as the conservation of energy. The second law is the most famous, and is often paraphrased as, "there is no such thing as a free lunch." Basically, it says, that every time you use energy to do work, you lose some of the energy forever. It becomes heat and radiates away, and you can never recapture it.

See HEAT; THERMODYNAMICS.

15. The Law of Biogenesis: Life Arises Only from Life

Until two Italian scientists set the record straight in the seventeenth and eighteenth centuries, many people believed in the spontaneous generation of life. Heaps of manure seemed to give rise to worms, and rotting garbage caused maggots to come into existence. Francisco Redi, working in the 1600s, showed that if meat were protected from adult flies trying to lay eggs, no maggots would develop. In the next century Lorenzo Spallanzani did similar experiments to show the same thing for bacteria. Establishment of the law of biogenesis, which may seem trivial today, was a major, necessary step in thinking toward a theory of evolution, which works only because life is transmitted from one generation to the next.

See BIOGENESIS; REDI; SPALLANZANI.

16. Cell Theory:
The Fundamental Unit of Life Is the Cell

Although the term "cell" (as referring to microscopic structures) was coined in the 1600s, it was not until 1839 that these structures were appreciated as the smallest units of an organism that can be said to possess life. Until Theodor Schwann and Mathias Schleiden, two German biologists, put forth their ideas in that year, most scientists thought that the units of life were fibers and vessels that started small and just grew, rather like crystals. Rudolf Virchow, another German, soon modified the cell theory to include the view that growth is the result of existing cells duplicating themselves. Virchow's modification was a key step that led to an understanding of how organisms develop from a single, unspecialized cell—the fertilized egg—to billions of specialized cells.

See CELL THEORY; SCHWANN; SCHLEIDEN; VIRCHOW.

17. Germ Theory: Microbes Cause Disease

Barely a century ago, Robert Koch, a German bacteriologist, became the first person to implicate a specific microbe—a bacterium—as the cause of a specific disease, anthrax, which afflicts hoofed animals. Koch went on to identify the bacterial species that causes tuberculosis. In so doing, he helped sweep away belief in evil spirits and other supernatural phenomena as causes of disease. The germ theory, as Koch's ideas have come to be called, led to a host of highly effective methods of dealing with much human suffering, from measures for preventing disease and ways to eliminate sources of bacteria to vaccines that render people immune and drugs that kill invading microbes.

See GERM THEORY; KOCH; JENNER (who developed the first vaccine).

18. Darwin's Theory of Evolution

The epochal contribution to science of Charles Darwin, the nineteenth-century English naturalist, should be divided into two parts. The first was his assertion that evolution had occurred; that the kinds of animals and plants we see today have not always existed, but have developed, through an accumulation of many small changes from remote ancestors. Others had suggested this in various forms, all the way back to the Greeks, but it was Darwin who marshaled the evidence finally to convince the world at large that it was true.

Darwin's second contribution was to suggest how the change happened. He pinned it all on what he called natural selection, which operates through four steps. First, organisms produce more offspring than can survive. Second, the offspring vary slightly in body and behavior. Third, the competition for food and mates will be won by the offspring with the more advantageous traits. And fourth, the winners will produce more offspring than will the losers, and so will pass those traits to a greater number of descendants. Darwin said that countless generations of such gradual change had caused one (or a few) original types of organisms to diversify into today's many forms. Some evolutionists today doubt that natural selection accounts for all evolutionary change. They hold that most species are not changing much during most of their existence; that major change can occur quickly in relatively few generations; and that random events can favor the survival of changed forms of a species as readily as can natural selection.

See EVOLUTION; NATURAL SELECTION; DARWIN; MODERN SYNTHESIS (which merged Darwin with Mendelian genetics).

19. Mendel's Particle Theory of Inheritance

Few major advances in science have been the work of one person, but such is the case with the discoveries of Gregor Mendel, the nineteenth-century Austrian monk who found that hereditary traits are passed on through discrete particles. Mendel raised sweet peas in his monastery garden, crossbreeding different varieties, and keeping records of the traits that showed up in the offspring. Through a statistical analysis (among the first in all of science) of the likelihood of a given parental trait showing up in an offspring, Mendel deduced that hereditary factors (later called genes) come in pairs, with one parent contributing each member of the offspring's pairs. He found that organisms may carry genes for traits they themselves do not show but which may be passed on to offspring. Mendelian inheritance patterns turn out not to be the only kind, but Mendel's discovery (neglected for forty years and only rediscovered around 1900) was correct as far as it went, and opened the door to modern genetics research that continues to develop at an ever-quickening pace.

See MENDEL; GENE.

20. Genes Are Made of DNA and Carried on Chromosomes

Gregor Mendel had shown that genes exist, but actually finding them in a cell and knowing what they were took some forty years of fairly intensive research after the rediscovery of Mendel's work. Chromosomes (the word means "colored bodies") had long been known to exist as structures in the nuclei of cells, but nobody knew what they did until Walter S. Sutton, an American biologist, noticed around 1902 that they behaved rather like Mendel's genes. For example, they came in pairs that split up when the organism made sperm or eggs, and then formed new pairs when a sperm and an egg united to create the offspring. In the 1920s Thomas Hunt Morgan, another American, found that various traits were inherited in predictable groups, and his analysis showed these traits had to be carried in linear fashion on chromosomes. In the 1940s Oswald Avery, still another American, finally figured out what kind of molecule it was that constituted genes. It was DNA (deoxyribonucleic acid), an enormously long molecule that was made up of many much smaller links, each being one of only four different kinds of simpler molecules. If DNA contained the code of life, its four-letter alphabet hardly seemed adequate to spell out all the specifications for a living organism.

See DNA; DNA, HOW ITS ROLE WAS DISCOVERED; GENES, HOW THEY WORK.

21. The Double Helix: How DNA Can Replicate

Perhaps the most famous discovery in modern biology was that of the molecular structure of DNA. In 1953 James Watson, an American, and Francis Crick, an Englishman, figured it out. Armed with only general knowledge, such as the relative amounts of the smaller molecules that are chained together to make DNA, the two played around with models made of balls and sticks and drawings on paper. The basic rules of chemistry dictated how long the stick, or chemical bond, should be between certain balls, or atoms, and what the angles would be between different sticks. When Watson and Crick hit on the fact that the DNA molecule consists of a double helix—two, single corkscrew-like strands entwined with one another—the structure immediately revealed how DNA duplicates itself so that a dividing cell can bequeath two identical sets of chromosomes. One of the most astonishing natural phenomena in all of chemistry, the self-replication of DNA, turned out to have an elegantly simple explanation.

See DOUBLE HELIX; DNA; WATSON; CRICK.

22. Cracking the Genetic Code: How Genes Work

The discovery of the DNA double helix did not explain what genes do or how they do it. In fact, the function of genes was worked out in the 1940s, a decade before the double helix, by two American biologists, George Beadle and Edward Tatum. In experiments with fruit flies and bread mold, they established that genes command cells to make specific chemicals, called proteins, of which there are thousands of kinds. They showed that each gene contains the genetic code for one kind of protein. Other researchers showed that each such code, preserved on the DNA in the cells' nucleus, was first transcribed into a similar molecule called RNA, which moves out of the nucleus. In the main body of the cell, special structures called ribosomes "read" the RNA code and accordingly assemble the proper protein molecule.

In the early 1960s, Francis Crick, who helped discover the double helix, and Sidney Brenner, a British molecular biologist, made the discovery that, more than the double helix, launched the modern era of molecular biology. They figured out how the seemingly simple four-letter alphabet of DNA could specify the proper sequence of protein subunits, called amino acids (which come in twenty different kinds) to make up a specific kind of protein molecule. The answer: a group of three "letters" specifies one amino acid. Armed with this knowledge, other workers quickly deciphered the three-letter codes for each of the twenty amino acids, and by 1966, the genetic code had been completely cracked.

One of the most profound discoveries made during this period was that the code is exactly the same for all plants and animals. Although minor exceptions—slight variations in a few microbes—have been found recently, this finding not only vindicated the reliance on insects and molds as clues to human heredity, but provided powerful confirmation of Darwin's theory of evolution. Had all life not stemmed from one ancestor, we would expect to find many genetic codes.

See GENES, HOW THEY WORK; DNA; RNA; RIBOSOME.

23. The Modern Synthesis: Unifying Darwin and Mendel

Charles Darwin in England and Gregor Mendel in Austria were contemporaries who made their chief contributions in the mid-nineteenth century. Darwin showed that the diverse species on Earth had all descended from a common ancestor by acquiring various modifications in body plan and form. But he had no idea how those modifications were transmitted from parent to offspring. Mendel, on the other hand, showed how differences in body plan and form are transmitted across the generations—through the discrete particles that would someday be called genes. But Mendel had no idea what role these particles played in the larger drama of the evolution of life. There is no evidence that the two scientists ever knew of one another.

Not until 1900, after both men were dead, was Mendel's long-ignored work rediscovered. At first it seemed to conflict with certain elements of Darwinian theory but, beginning in the 1920s, a number of biologists saw how to reconcile the discoveries of both men. The reconciliation was called the "modern synthesis." Mendel's genes, everyone soon realized, were the means by which Darwin's modifications could be passed on.

See MODERN SYNTHESIS; MENDEL; DARWIN; EVOLUTION THEORY.

24. The Evolution of Human Beings

A fitting climax to this sequence of science's greatest achievements—at the opposite end of the spectrum from the origin of everything in the Big Bang—would seem to be the modern understanding of how human beings originated. The appearance of our species in the world is, after all, a relatively recent event, having occurred only in the latest 0.01 percent of the time since the universe began.

Despite controversy among the paleoanthropologists who study this question, the story of human evolution is remarkably clear. We are, as Darwin said, descended from the animals, from small African primates that probably looked a little like modern apes but which weren't apes. (Apes descended from the same ancestors, changing as much as humans have since the two lineages parted. Thus it is as true to say that apes descended from human-like ancestors as vice versa.) By three and a half million years ago, we were walking on two legs but carrying brains the size of those in modern chimpanzees. By two million years ago, our bodies were almost completely modern in size and appearance, and our brains had grown to half their modern size. Our faces had lost much of their muzzle and our hands had found an ability to chip stones into tools. By one million years ago we had migrated out of Africa and through much of Asia and Europe. We had domesticated fire and made clothes to protect us in Eurasian winters. By 100,000 years ago our brains had reached fully modern size. Further evolution in brain and body produced only minor revisions until, by 40,000 years ago, fully modern human beings were on the scene.

Not until about 2,500 years ago did human beings begin the enterprise called science. It had only a brief life, however, and died out within a few centuries. Fortunately, human beings revived the practice of science about 500 years ago.

See HUMAN EVOLUTION.

PART 3

Alphabetical Entries

ABSOLUTE ZERO—This is the lowest possible temperature that can be achieved. Since heat is actually the vibration of atoms within a substance (the faster the atoms vibrate, the hotter it is), absolute zero is a temperature so cold that the atoms are no longer moving. On the Fahrenheit scale it is 459.69 degrees below zero. That's 273.16 degrees below zero on the Celsius scale. There is also a "Kelvin scale" on which absolute zero is 0 degrees. In practical terms, the coldest anyone has been able to chill anything is to within a fraction of a degree of absolute zero.

See TEMPERATURE; HEAT; KELVIN.

ACCELERATOR—A common shorthand term for PARTICLE ACCELERATOR.

ALTRUISM—This is the sociobiologist's term for instinctive behavior in which an organism uses some of its own energy or risks its own life for the benefit of another organism. Parental care of the young is an obvious example. Less obvious is the habit of some animals to sound an alarm call when they spot a predator, thus drawing the predator's attention to themselves. Bees, which commit suicide when they sting an intruder into the hive, demonstrate ultimate altruism. Such behavior has long posed an evolutionary puzzle: How could an instinct that *reduces* one's chances of survival persist? After all, natural selection favors instincts that improve survival. Sociobiology's answer is that such behaviors are exercised to favor the survival not of the individual but of close relatives carrying the same genes. If the sacrifice of one individual (carrying one set of altruistic behavior genes) enables several individuals (carrying the same genes) to survive when they would not otherwise have done so, the behavior will spread. The animals never have to know a thing about genetics. All that matters is that the benefit goes to more animals carrying the same genes than are lost in the exercise of the sacrifice.

See SOCIOBIOLOGY; NATURAL SELECTION.

AMINO ACIDS—These are best known as the building blocks of protein molecules. All living cells synthesize protein molecules by linking amino acid molecules into long chains that then fold up into characteristic shapes. A protein molecule may be made of several hundred or several thousand amino acids. The exact number and sequence of the amino acids in a protein is specified by the genes of the cell that makes the protein. There are some 200 different kinds of amino acids known, of which about 80 are commonly found in living organisms. Of these, just 20 kinds are used to make proteins. Amino acids are made up of various combinations and numbers of atoms of carbon, hydrogen, oxygen, and nitrogen.

See PROTEIN; GENES, HOW THEY WORK.

ANTIBODY—Antibodies are special protein molecules that circulate in the blood and which have a shape that causes them to bind to specific foreign protein molecules (called antigens) that invade the body. Antibodies are an important component of the body's system for resisting attack by disease-causing bacteria and viruses. Since all such invading microbes are at least partly made of protein molecules, antibodies can bind with these molecules, disabling the microbe and in effect holding it until white blood cells (which live rather like amoebas) consume it. The body's immune system can manufacture thousands of different kinds of antibodies, each tailored to bind to a specific invading protein. When first infected by a particular kind of microbe, the body manufactures new antibodies to combat the infecting microbes and then keeps a supply of the same antibodies on hand, sometimes for many years, to block reinfection by the same kind of microbes.

ANTIGEN—Any protein molecule that can get inside the body and stimulate a response from the immune system, the body's natural defense mechanism. Antigens commonly enter the body on the surface of bacteria, viruses, and pollens. The body fights back by making antibodies. See ANTIBODIES.

ANTIMATTER—Although the existence of antimatter is a favorite topic of particle physicists, the fact is that the only known examples of antimatter are subatomic particles created in the laboratory, in particle accelerators that slam subatomic particles together, briefly converting energy into particles having mass. It routinely happens that these processes create equal amounts of mass as matter and antimatter. Antimatter particles do not last long, however, for as soon as they encounter their equivalent in matter, the two mutually annihilate, converting both back into energy.

Antimatter particles created in the lab appear to be exactly like ordinary matter except that they have opposite electrical charges. As a result, you could, in principle, have an atom of anti-hydrogen with an anti-electron (called a positron) orbiting an anti-proton. The laws of physics say a sun made of anti-hydrogen would shine as brightly as our own sun.

If atom smashers produce equal amounts of matter and

antimatter, many physicists think that the Big Bang, the explosion that began our universe, should have done the same thing. Since the two annihilate one another, the universe should then have disintegrated back into pure energy almost as soon as it formed. That it obviously didn't has forced physicists to wonder whether the Big Bang somehow produced an imbalance of the two forms such that, even after mutual annihilations, there was enough matter left over to make the universe we know today. Furthermore, all evidence suggests that the other stars in our galaxy are stars and not anti-stars, and there is reason to believe that most or all other galaxies are just that, and not anti-galaxies, even though matter and antimatter should have been balanced. Physicists find such asymmetries ugly and out of keeping with a universe in which virtually every other phenomenon behaves with unfailing symmetry. It's a puzzle and scientists still can't figure it out.

See MATTER; ATOMIC STRUCTURE; PARTICLE ACCELERATOR; BIG BANG.

ARCHIMEDES, 287 B.C.–212 B.C., Greek mathematician—Archimedes is the ancient wiseman who supposedly leaped from his bathtub, having suddenly grasped the solution to a problem, and ran naked through the streets of the ancient Greek city-state of Syracuse in Sicily shouting, "Eureka." The story is of dubious authenticity, but Archimedes definitely did find the solution to the problem in the legend—how to tell whether a king's crown was pure gold or not—and to a host of other problems from theoretical mathematics to the invention of fearsome war machines.

Archimedes was born in Syracuse. His father was an astronomer, and he may have been related to Syracuse's King Hieron II. Archimedes studied at the great academy in Alexandria, learning the geometry created by Euclid (who lived a century earlier). After returning to Syracuse, he went on to surpass Euclid as a mathematician.

Archimedes was, in fact, the greatest ancient mathematician and scientist that we know of. He was the founder of theoretical mechanics, and some think he was not equaled in mathematics until the time of Isaac Newton, nearly 2,000 years later.

Archimedes developed a host of solutions to geometric and mechanical problems, such as calculating areas and volumes and measuring the forces multiplied by levers. Algebra had not been invented yet, so his statements of formulas were in ordinary language. For example, "The surface of a sphere is equal to four times the area of the greatest circle in it." When algebra was invented much later, his succinct words were reduced to $A = 4\pi r^2$. He also worked out the laws of mechanics that describe, for example, how forces can be multiplied with levers and pulleys. In fact, he invented the compound pulley, a device that makes it possible to lift very heavy weights with much lighter pulls. Archimedes' prowess with such devices led him to say, at least as legend has it, "Give me a place to stand on, and I will move the earth."

Many of Archimedes' most famous inventions were practical applications of the mechanical laws he discovered. For the king he designed war machines such as huge catapults for hurling stones at the Roman soldiers who beseiged Syracuse, and in another doubtful story, arrays of mirrors that focused sunlight on Roman ships, setting them afire. He also invented a screw-like device to raise water for irrigation—the so-called "Archimedes screw" still in use in some parts of the world.

The famous problem with the king's crown led to the establishment of what is known today as Archimedes' principle. It seems that King Hieron had ordered a crown made of pure gold. He wanted to be sure that the goldsmith had not cheated him by alloying some silver with the gold. Everybody knew that silver was lighter than gold and that the crown weighed what it was supposed to. It was also obvious that a devious smith could simply have added extra silver to bring the crown up to the right weight. If so, the crown would have a larger volume than pure gold of the same weight. The problem was how to measure the volume of an irregular object. Archimedes

puzzled over this until he happened to step into a full tub
and notice that the water that spilled equaled the volume
of his own body below the surface. He realized that he
could submerge the crown and measure the volume of
water displaced. If it was more than the amount displaced
by an equal weight of gold (which being heavier would
have a smaller volume), the king could accuse the goldsmith.

From this observation Archimedes went on to discern
the principle named for him: than an object immersed in
water is buoyed up by a force equal to the weight of the
water displaced, and the corollary that a floating object
displaces a volume of water that weighs as much as the
object.

Despite Archimedes' war machines, King Hieron's
forces finally yielded to the beseiging Romans. When they
finally broke into the city, a pillaging Roman soldier happened upon Archimedes, then seventy-five years old and
intently scratching geometric diagrams in the dust. The
soldier called to the old man but Archimedes did not look
up. Angered, the soldier killed Archimedes with a sword.

ARISTOTLE, 384 B.C.–322 B.C., Greek philosopher and
scientist—Over the long history of Western civilization,
probably no other person has been cited more often as an
authority on almost every subject than Aristotle. Unfortunately, despite his enormous influence, Aristotle was
wrong so often that, on balance, it could be argued that he
did more to delay the progress of science than to advance
it. Aristotle delved into every area of knowledge available
in his day, from mechanics and chemistry to embryology
and cosmology—all in addition to the purely philosophical
matters for which he was also famous. His intellectual system was the basis for medieval Christian and Islamic scho-
lastic philosophies. His scientific treatises were quoted as
the last word on many subjects through the middle ages,
and in some cases up until the scientific revolution of the
sixteenth and seventeenth centuries—a scientific hege-
mony of some 2,000 years.

"If I had my way," Roger Bacon declared in the thirteenth century as he strove to encourage a new approach to knowledge, "I should burn all the books of Aristotle, for the study of them can lead to a loss of time, produce error, and increase ignorance."

Aristotle held that the Earth was at the center of the universe, that all matter was made of four elements (earth, air, fire, and water), that heavy objects fall faster than light ones, that breathing exists to cool the body, that the notion of atoms (which Democritus, another Greek, was advocating) was wrong, and that mammals' hearts have three chambers (four is correct). He was wrong on these and many other points—especially in the physical sciences—but in fairness, he was right about a few things, especially in the biological sciences. For example, he overthrew the belief that the sex of an embryo is determined by its site of attachment in the womb. He also argued against the idea that embryos are simply miniature adults, with all parts preformed.

The main reason Aristotle did so much better in biology than in physics was that he took the trouble to examine biological phenomena—he poked about seashores and dissected animals—whereas on physical matters he relied almost entirely on logic alone. For example, when dealing with the question of how fast objects fall, he tested a leaf and a stone and concluded that objects of greater weight fell faster than those of lighter weight. It was a real experiment, to be sure, but from this point on, Aristotle relied on his mental faculties alone. It was therefore logical, he concluded, that a two-pound weight would fall twice as fast as a one-pound weight. This latter point he never bothered to test with an experiment, but his verdict was solemnly repeated for nearly 2,000 years until Galileo actually went out and did the experiment to show that, where wind resistance is not a complicating factor, all objects fall at the same speed.

Aristotle's reliance on the powers of the mind alone to solve problems may have been his greatest disservice to science. The scientific revolution, when it came, would emphasize the need for close observation of nature and experimentation. Observed facts, later generations would learn, always take precedence over theory, no matter how beautiful the logic behind it. For Aristotle, on the other hand, there was no distinction between science as we know it today and the purely mental practice of philosophy. His paramount goal was the cultivation of the mind, and his belief was that truth could be perceived by the exercise of the developed powers of the mind.

Aristotle was born the son of a physician who served in the royal court of Greek Macedonia. At the age of seventeen, his parents both dead, he was sent by his guardian to Plato's academy in Athens. There for the next twenty years he remained, engaging in intellectual dialogues with Socrates and Plato—one of the greatest intellectual concentrations the world has seen. When Plato died and Aristotle ·was not appointed to succeed him as head of the academy, the Macedonian scholar left Athens with a close circle of friends, and traveled for the next twelve years throughout Greece and Asia Minor. During this period Aristotle saw fit to prescribe the ideal ages for marriage—thirty-seven for men and eighteen for women. As it happens, Aristotle himself married at the age of thirty-seven, and scholars surmise that his wife must have been around eighteen.

At the age of forty-two, Aristotle was invited by the king of Macedonia, Philip II, to tutor his thirteen-year-old son, who would one day be known as Alexander the Great. He did, and years later, when Aristotle published certain treatises, Alexander became angry when he realized that the knowledge vouchsafed to him was not to remain his alone.

At the age of fifty Aristotle returned to Athens and established his own school, called the Lyceum, which specialized in science and history. He undertook his many investigations of plants and animals and received exotic specimens sent to him by traveling associates. Some apparently came from Alexander as his conquering army swept through much of the ancient world.

Perhaps Aristotle's greatest contribution to science was in the form of recognizing that there is order in the living world. Plants and animals do not have a random jumble of traits, but share many features and may be classified on the basis of what they share—numbers of legs, methods of reproduction, and so forth. Aristotle's perceptions of different levels of complexity among organisms led to the idea of a "scale of nature" from the simplest at the bottom to the most complex—the human being—at the top. Although modern evolutionists would resent some of the implications of this idea, it had great value in the effort to make sense out of nature.

At the age of sixty-one or sixty-two, Aristotle faced an outbreak of anti-Macedonian agitation in Athens when Alexander died. He was indicted on a charge of impiety for having written a poem twenty years earlier that virtually deified a slain king who was also a friend. The deification of a mortal was a capital offense, but before Aristotle could be taken prisoner, he escaped to another city-state, where he died of a stomach ailment the following year.

ATOM—Every form of matter in the everyday world is made of atoms. There are ninety-two kinds of atoms in the natural world and, depending on which ones you choose and how you put them together, you can make anything that exists. Atoms consist of a nucleus, made of protons and neutrons, around which "orbits" a cloud of electrons. The neutrons and protons are, in turn, made of quarks, three in each type of particle. A major question in modern physics is whether quarks are themselves made of still smaller particles. Beyond the everyday world—inside stars, for example—matter may exist as separate neutrons, protons, and electrons.

The factors that determine an atom's chemical properties are the number of protons in its nucleus and how this number relates to the number of electrons in orbit around the nucleus. If the numbers are equal, the atom does not combine with other atoms. If there are more protons than electrons, the atom will tend to combine with an atom that has the opposite problem, more electrons than protons. The atom's goal, if it can be called that, is to try to balance the numbers of its protons and electrons. Combinations of atoms make a molecule. Thus, a molecule of water is made of two atoms of hydrogen bound to one atom of oxygen. Many kinds of molecules are much more complex than this. A single protein molecule, for example, may be made of tens of thousands of atoms.

See ATOM THEORY (the discovery that atoms exist at all); ATOMIC STRUCTURE (the world inside the atom); CHEMICAL BONDING, THE NATURE OF.

ATOMIC STRUCTURE—For most practical purposes, the internal structure of an atom can be understood rather easily. All atoms consist of a nucleus surrounded by a roughly spherical shell in which one or more electrons is orbiting. Hydrogen, the simplest atom, has a nucleus that consists of one particle called a proton. Orbiting this lone proton is a single electron. Atoms of all other elements have the same basic form except for two things: (1) Their nuclei contain several protons, along with about an equal number of other particles called neutrons. (2) For each additional proton in the nucleus, the atom contains an equal number of electrons in orbit.

And that, at one level of understanding, is it. The atom consists of just three kinds of particles—protons and neutrons clumped at the center and electrons whizzing about in essentially concentric shells, rather like the layers of an onion. In truth, electrons do not follow strictly circular orbits but, instead, wander crazily in clouds surrounding the nucleus. Nearly all the mass of an atom is in the nucleus. Protons and neutrons weigh about 1,800 times as much as electrons.

Before probing the innards of atoms more deeply, one should first appreciate their size. They are small; very small. One drop of water contains more than 100 billion billion atoms or, in scientific notation, 10^{20} atoms. If an atom were enlarged to the size of a drop of water, a drop of water enlarged to the same extent would be about as

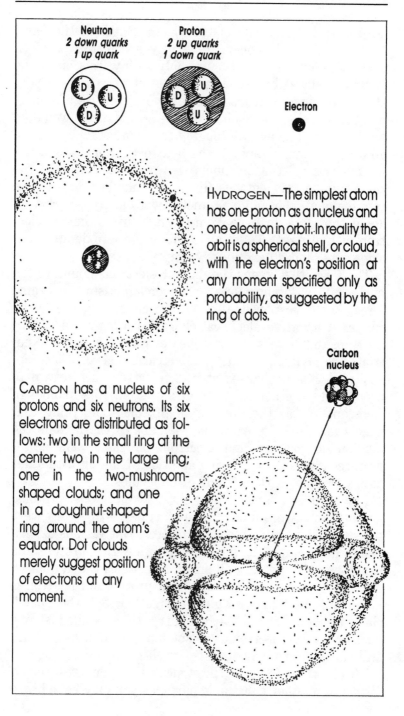

Neutron
2 down quarks
1 up quark

Proton
2 up quarks
1 down quark

Electron

HYDROGEN—The simplest atom has one proton as a nucleus and one electron in orbit. In reality the orbit is a spherical shell, or cloud, with the electron's position at any moment specified only as probability, as suggested by the ring of dots.

Carbon nucleus

CARBON has a nucleus of six protons and six neutrons. Its six electrons are distributed as follows: two in the small ring at the center; two in the large ring; one in the two-mushroom-shaped clouds; and one in a doughnut-shaped ring around the atom's equator. Dot clouds merely suggest position of electrons at any moment.

large as all the oceans of the Earth. Small as they are, atoms still mostly consist of empty space; the particles that make up an atom occupy only a tiny fraction of the volume enclosed by the outermost electrons. If the proton at the center of a hydrogen atom were the size of a marble, the one electron orbiting it (the size of a BB) would be about half a mile away. Roughly the same scale applies to all other atoms. Uranium, the heaviest natural atom, has a nucleus consisting of ninety-two protons and about the same number of neutrons (the exact number depends on the isotope). Uranium's ninety-two electrons orbit at about the same distance from the nucleus as the one electron of hydrogen. Thus, all atoms are approximately the same size, and all are mostly empty space.

The force that keeps the distant electrons bound to the nucleus is a familiar one—electromagnetism, the same force that is responsible for "static cling." Each electron carries a negative electrical charge. Each proton carries an equal but opposite (positive) electrical charge. (Neutrons are neutral, having no electrical charge.) Since opposite electrical charges attract, electrons are drawn to protons. In the normal form of any atom, the number of electrons equals the number of protons. The thing that makes atoms differ from one element to the next is the number of protons and electrons in them. These differences give atoms differing abilities to link up with other atoms to form various kinds of molecules.

For a while scientists thought that electrons, protons, and neutrons were fundamental particles—that they were the smallest building blocks of matter. But various experiments have led to the fairly recent view that protons and neutrons are made of still smaller particles called quarks. According to current theory, each proton and neutron is made of three quarks "glued" together by other particles—dubbed "gluons"—that carry the so-called strong nuclear force. Electrons are not made of quarks and are still considered fundamental particles.

This is where ideas on the structure of atoms rest today. Most of the other subatomic particles you may have heard about—the positron, the many kinds of mesons, and others—do not exist in the ordinary world. They are created inside particle accelerators when atoms are smashed together at enormous velocities. The energy released by such collisions causes quarks to combine in odd ways that are unstable, and which break apart after only a tiny fraction of a second. Also, the energy imparted by the accelerator can itself be converted into new, short-lived particles, according to the teaching of Einstein's relativity theory that mass can become energy and energy can become mass.

The only known stable particles besides atoms are photons and neutrinos. Photons are tiny bundles of electromagnetic energy that constitute light, radio waves, and all other kinds of waves in the electromagnetic spectrum. The sun floods the Earth with photons. Within the atom, the electromagnetic force that binds electrons to protons is carried by photons that continually move back and forth between these oppositely charged particles. Neutrinos are particles with no charge and, apparently, little or no mass. They are produced when certain kinds of particles decay, but they then simply whiz off through space, never to interact again with other particles. As it does with photons, the sun floods the Earth with neutrinos, but they are so small and so unresponsive to forces that they pass straight through the Earth without doing anything.

This, as was said at the outset, is a view of the atom that suffices for most practical purposes. Physicists, however, like to probe matters a little deeper. They would like to understand, for example, how it is that matter and energy can alternate with one another. If you take a loose proton and push it through a particle accelerator until it is moving at nearly the speed of light, and then smash it into another particle, the two particles not only break up into smaller pieces (flying off in all directions), but the energy of the two original particles' motion is converted into new particles. Quite literally, matter is created. Experiments of this kind have led physicists to realize that the phenomena that produce this new matter lie behind the existence of all matter. All the protons, neutrons, and electrons that make up this book, your body, and everything else are actually quantities of energy packaged in ways that lead us to interpret them as something other than energy—as matter.

Common sense and the English language are reasonably good guides to understanding most areas of science, but they seem to fail when it comes to particle physics. Physicists have even found that particles of matter can pop into existence without the use of particle accelerators. It happens all the time in nature. Energy that is, in a sense, just floating around can suddenly congeal into a speck of matter, and just as suddenly that matter can pop back into a burst of energy. Furthermore, certain combinations of protons, neutrons, and electrons are inherently unstable, and atoms made of these combinations will spontaneously transform themselves into more stable atoms, converting the ill-fitting particles into energy. The process is called radioactive decay. The most common form of uranium, for example, decays through a series of fifteen steps, at each of which a burst of energy is emitted, before it reaches a stable form which, because eighty-two protons are left in the nucleus, is lead.

In their study of atomic structure, particle physicists have come face to face with the most profound questions philosophers have ever raised: What is the fundamental nature of existence?, and How did the world begin, right down to the atomic level? Modern physics is very much wrapped up with such questions, and matters related to them are already the subjects of many books.

See PROTON; NEUTRON; ELECTRON; QUARK; PHOTON; NEUTRINO; ELECTROMAGNETISM; STRONG FORCE; QUANTUM THEORY (which deals with the fact of energy existing only in packets); ANTIMATTER (which holds that for each particle there is an antiparticle); GRAND UNIFIED THEORY (which seeks to explain all particles and forces as derived from a single primordial superforce); BIG BANG (the event that created particles and forces).

Also see biographical sketches that give the history of the discovery of atomic structure: THOMSON (who discovered the electron); RUTHERFORD (who suggested that the atom had a positive nucleus surrounded by a swarm of electrons); BOHR (who developed the quantum theory of the atom); PLANCK (quantum mechanics); EINSTEIN (relativity theory); HEISENBERG (uncertainty principle).

ATOM SMASHER—A colloquial term for a machine that physicists use to hurl substantive particles at a target. When the particles "smash" the target, the energy released in the collision transforms itself into new subatomic particles. This is how physicists find out what atoms are made of. See PARTICLE ACCELERATOR.

ATOM THEORY—Although the ancient Greeks gave us the term "atom," it was not until the twentieth century that scientists finally clinched the proof that all matter really was made of tiny particles. Einstein was among those scientists but, ironically, he also led the way to showing that the Greek word was unfortunately not quite apt.

Atomos is the Greek word meaning "undivided." (Einstein, of course, helped show that atoms could be divided.) The word appears to have been first used in this sense in the fifth century B.C. by a philosopher called Leucippus. Nothing much is known of Leucippus except that his ideas were taken up by another Greek philosopher named Democritus, who usually gets misplaced credit for originating the atom theory. "In reality," Democritus said, "there are only atoms and the void." It still makes a fairly succinct statement of the nature of reality.

The ancients, including the Roman poet Lucretius in the first century B.C., developed the idea to a surprisingly sophisticated level. First, they said that if you could cut up any object into its smallest pieces, you would eventually come to an ultimate particle that could not be further divided—the atom. Second, the ancients claimed there were many kinds of atoms of various sizes and shapes, and that depending on how they were combined, you could produce all the physical properties of matter in the resulting bulk product. And third, they thought that all the known phenomena of the physical world were the result of combinations of forces exerted by atoms as they moved and collided.

They didn't get the third point quite right, but ancient atomic theory was a stunning intellectual achievement. Nonetheless, it lapsed into obscurity for at least 1,600 years. Not until the time of Galileo, Newton, Boyle, and Hooke did the concept of atoms re-emerge. These later researchers believed in atoms, and Hooke even went considerably ahead of his time by suggesting that the properties of gases could be explained as the motion and collision of clouds of floating atoms.

The clearest statement of atom theory, however, came in 1808 with John Dalton's publication of "A New System of Chemical Philosophy." Dalton was an English chemist, and his system quickly became the foundation of modern chemistry and, more specifically, of modern atomic theory. Dalton put forth several ideas that remain central to the modern understanding of the nature of matter. First, he said that each chemical element was made up of its own special atoms, there being a different kind of atom for each element. Then he said that atoms of a given element are identical in weight, shape and all other characteristics. Dalton held that chemistry consisted simply of linking atoms into what he called compound atoms (and which we call molecules), or of unlinking atoms. He insisted that atoms were never destroyed or changed, but only switched from one combination to another.

Dalton arrived at his ideas through his interest in weather, and specifically the air. By Dalton's time other scientists had established that air was not a single gas but a mixture of several gases, mainly nitrogen, oxygen, carbon dioxide, and water vapor. Dalton collected air samples from many places and analyzed each for its composition. He was puzzled by the fact that carbon dioxide, the heaviest known gas in air, didn't sink to the bottom of the atmosphere; air from mountaintops had just about as much of it as air from valley floors. The only way Dalton could explain this was by conceiving of each gas as having its

own kind of atom, and that the atoms were so different
that they essentially ignored one another. Carbon dioxide
was evenly distributed through the atmosphere because it
behaved independently of the other gases. Dalton assumed that atoms must be so small that those in air are
relatively far apart, and seldom bump into one another as
they whiz about.

An important part of Dalton's theory came also from his
work in analyzing the weights of various chemicals that
were consumed in a chemical reaction. Much of Dalton's
data came from the knowledge of chemists and manufac-
turers who, by trial and error, had learned that if, for
example, you wanted to make 100 pounds of a given compound you needed 33 pounds of each of three ingredients.
In such data, Dalton perceived certain regularities. Chief
among these was that substances always had predictable
combining ratios. In other words, if matter came in tiny
particles, the particles always linked up in a regular fashion. Consider, for example, that you are in the business of
making water, and you invariably find that it always takes
twice as much hydrogen as oxygen. If you were Dalton
and already thinking that oxygen (like the other components of air) always comes in identical tiny particles called
atoms, you could easily come to the view that water is
made by binding two hydrogen atoms to one oxygen atom.

Dalton's ideas did not immediately prevail. In fact, they
initiated a hundred years' war that did not see the end of
significant resistance to atom theory until the early 1900s.
Many scientists regarded the idea of atoms as a mere intellectual aid, a glorified algebraic symbol that helped
make chemistry understandable. The clinching arguments
came from the study of Brownian motion—the perpetual
dancing of small particles suspended in a liquid. Botanist
Robert Brown had first noticed this effect on pollen grains
in water in the 1820s, but others found that all sorts of
dustlike particles would refuse to settle to the bottom no
matter how long one waited. Nobody could explain the
phenomenon until nineteenth-century atomists proposed
that the particles were kept in their jerky motion because

they were repeatedly being bumped by particles of water. If water were a continuous material, there would be nothing to push a foreign particle around. The force exerted by the continuous material would be equal on all sides, and the particle, if heavier than water, should sink.

In 1905 Einstein published four articles that shook physics to its foundations, including one on relativity theory. One of the less well remembered articles presented a set of equations that could be used to prove the existence of atoms from measurements of how particles zigged and zagged during Brownian motion. Einstein lacked the means to measure the motion accurately, but shortly afterward the equations fell into the hands of a French physical chemist, Jean Perrin, who had been trying to measure Brownian motions for ten years. His calculations proved that the particles could move only if water existed in discrete particles.

Strictly speaking, Einstein and Perrin had proven only that matter existed in particles, but that was enough to settle the atomism debates, for they revolved not so much around the nature of the particles as on whether they existed at all. Although it remained for a new generation of scientists to determine the structure of the particles, that generation would in large measure confirm Dalton's views.

See DALTON; EINSTEIN; ATOMIC STRUCTURE; ELEMENT; MOLECULE.

AUSTRALOPITHECUS—This is the genus name of the earliest known human ancestor after the human lineage evolved away from the ape lineage. Its literal meaning is "southern ape," which made sense when the only known specimen was from South Africa. That meaning is now irrelevant, however, since *Australopithecus* remains have also been found in eastern Africa. There are several species within this genus.

See HUMAN EVOLUTION.

AVERY, OSWALD, 1877–1955, American biochemist—See DNA, HOW ITS ROLE WAS DISCOVERED.

BACON, FRANCIS, 1561–1626, English philosopher of science—Francis Bacon, a lawyer, statesman, and philosopher, was influential in the intellectual and philosophical life of his time, but was not much of a scientist. He is chiefly remembered for advocating a special brand of inductive reasoning as the best way of making intellectual progress. The method is hard to understand. Although Bacon insisted it was foolproof, it nonetheless failed to gain much of a following, and Bacon's own beliefs about the laws and actions of nature were very much out of date even in his own time. Bacon is also remembered for advocating the idea, novel in his day, that the purpose of science is to improve the lot of mankind. Bacon thought that if only the Crown would spend enough to support a few years of dedicated scientific investigation, just about everything of practical value could be learned.

Francis Bacon should not be confused with Roger Bacon, an English Franciscan friar of the thirteenth century, who was also a philosopher of science, but more important an experimenter, supplementing guesswork with evidence. (See below)

BACON, ROGER, 1220–1292, English philosopher, scientist—Roger Bacon was a Franciscan friar who developed an interest in science, advocating experimentation as the best way to learn the secrets of nature. Prior to his time, science had consisted mainly of accepting the teachings of Aristotle. Despite his views, Bacon did not do much experimenting himself, though he did some work with lenses and mirrors. He went on to describe eyeglasses, which soon thereafter came into use. He was the first person in the West to write accurately on how to make gunpowder, although it was not until the next century that guns were invented.

Bacon was also something of a futurist who imagined flying machines (they would flap their wings, he thought) and proposed mechanically propelled ships and land carriages.

Roger Bacon should not be confused with Francis

Bacon, who lived 300 years later and who advocated a special kind of inductive reasoning, rather than experimentation, as the way to make progress in science.

BACTERIA—These are the smallest and simplest of living organisms, so small that 500 average-sized ones lined up would equal the thickness of a dime. (Viruses, which are smaller, are really not full-fledged organisms.) Bacteria are neither plants nor animals, but are in a separate kingdom called Protista. They differ mainly from one-celled plants or animals in having a vastly simpler internal anatomy. The simplicity is most extreme in that bacteria have no discrete nuclei—the central repositories for plant and animal genes. Instead, bacterial DNA is scattered within the cell. The vast majority of bacterial species are harmless to people and carry on independent lives as the chief agents of organic decay. A few cause disease. (See GERM THEORY.) Some, such as *Escherichia coli*, a species that inhabits the human intestine, are needed to help digest food or perform other functions, but can cause disease if they get into the urinary tract or some other part of the body that they do not customarily inhabit. Despite being the simplest of organisms, bacteria have proven extremely useful in studies of genetics, since the genetic code is essentially the same for all organisms. Modified strains of natural bacteria are used in recombinant DNA technology to synthesize needed proteins. Human genes, spliced in among bacterial genes, work perfectly.

See GENETIC CODE; GENES, HOW THEY WORK.

BACTERIOPHAGE—Even germs have germs. A bacteriophage is a type of virus that infects bacteria. Like all viruses, bacteriophages lack the machinery to carry out the instructions coded in their genes. They must invade a truly living cell and commandeer its protein-making machinery to reproduce. Research on bacteriophages has proven enormously useful in learning how the work of genes is regulated.

See GENES, HOW THEY WORK.

BEADLE, GEORGE, 1903–   , American molecular ⟍ ogist—See DNA, HOW ITS ROLE WAS DISCOVERED.

BIG BANG, THE ORIGIN OF THE UNIVERSE—In 1929 a California astronomer named Edwin Hubble made a discovery that was to change forever our concept of the nature of the universe. It had been recognized for a few years that galaxies, vast star systems that contain virtually all known celestial objects, seemed to be flying away from Earth. Hubble found that the faster a galaxy was moving, the farther away it was. This relationship contained one of the most startling implications in all of science: That all galaxies are racing away from one central point, and that in the distant past they must therefore have all been very close together. The only way they could have reached their present positions and motions would be if some colossal explosion, generating energy beyond all human comprehension, had sent them flying away from that central point.

The explosion idea soon came to be known as the Big Bang theory, and for the next thirty-six years it challenged the older theory of a static universe, in which the galaxies are claimed to have always been where they seem to be today. At first many scientists resisted the Big Bang theory because it seemed to call for one highly extraordinary event—a supernatural event beyond the realm of science—to touch off the original explosion. Scientists tend not to like suggestions that the explanations for natural events may be off limits to them.

Some astronomers contrived an alternative theory, accepting that the universe was expanding, but without a cataclysmic beginning. The result was the Steady State theory, in which new matter—such as new stars and galaxies—was thought of as constantly being created to fill the gaps that grow as old matter flies apart.

In 1965, however, two scientists who worked for Bell Labs, Arno A. Penzias and Robert W. Wilson, discovered something that settled the debate. They detected and measured the energy left over from the earliest stage of the Big Bang. It showed up in their instruments as heat measuring only about 3 degrees above absolute zero, but it was there, detectable in every direction from the Earth. The nature of the primordial heat was undeniable evidence—in fact, virtual proof—of the Big Bang theory, and since that discovery, virtually all astronomers have had to accept the fact that the universe began with an event so out of the ordinary that it may never be understood. The universe as we know it has a finite age (between 13 and 20 billion years, depending on various methods of estimating), and before its birth there was not even such a thing as time.

The beginning of such unwieldy ideas—the finding that the universe is expanding—took place in the 1920s, when various astronomers studied the peculiar nature of the light coming from distant galaxies. At the time, nobody knew they were galaxies, or great clusters of billions of stars. They were simply seen as bright, fuzzy objects, and were called nebulae.

Still, the light coming from them was found to contain important information. When light from a star is spread out into a spectrum, much as a prism separates white light into its component colors, the spectrum contains gaps. The gaps, seen as dark bands, are caused when certain chemical elements in the light source absorb light of certain frequencies, or colors, emanating from the source. In experiments on Earth, for example, calcium has been found to absorb light from two parts of the spectrum. When a spectral analysis of starlight shows these same two "absorption lines," the star is presumed to have calcium in it. Various elements produce characteristic sets of absorption lines that act as "signatures" of their presence in a star or other matter.

When the distant nebulae were subjected to spectral analysis, various signatures showed up, but in the "wrong" portion of the spectrum. The absorption lines were invariably shifted toward the red, or lower frequency, end of the spectrum. The only plausible explanation for this "red shift," physicists quickly realized in the

1920s, was a phenomenon known as the Doppler effect, which is familiar to most people in the way it alters the frequency of sound when the source and the listener are moving rapidly with respect to one another. When, for example, a speeding train blows its whistle as it approaches, we hear it as a particular pitch, or frequency. But once the train passes and is moving away, the tone quickly shifts to a lower frequency. A passenger on the train might hear the effect as it alters the tone of a bell at a crossing. (It does not matter whether the source or the listener is moving, only that the distance between the two is changing. For a fuller explanation, see the entry on DOPPLER EFFECT.)

Since different galaxies showed different degrees of red shifting of the spectral absorption lines for certain elements, the clear implication was that each was moving away from the Earth at a different speed. The fastest objects in the universe are estimated to be moving away from us at about 90 percent of the speed of light, or about 167,000 miles per second.

Hubble's chief contribution was in recognizing that the various degrees of red shift were correlated with the distances of the galaxies. This phenomenon can be explained by analogy with a loaf of raisin bread. Before the dough is allowed to rise, the raisins are distributed throughout it at various distances from one another. As the dough rises and then is baked, the loaf expands at a constant rate throughout. An observer on raisin A might see raisin B move from an inch away at the start to two inches at the end. In the same time, however, raisin C, which started four inches away from A, would also have doubled its distance, moving to eight inches away. Simple arithmetic would show that C was moving four times as fast as B. If raisins had red shifts, C would show more of a red shift than B. Hubble could not measure the distances between galaxies, but the red shifts told him that their pattern of movement had to be like that of the raisins.

The raisin-bread analogy also helps explain a common popular misconception about the expanding universe. If all the other galaxies are moving away from our own, it might seem that we are in the center of the expanding universe. The fact is that an observer on any raisin in the dough would see the same effect. The distance between any one raisin and all the others grows, and does so at the same rate. Thus, while we see the most distant celestial bodies flying away from us at 90 percent of the speed of light, observers on those bodies would say that we in the Milky Way galaxy are moving away from them at 90 percent of the speed of light.

Another common misconception is the idea that galaxies themselves are expanding. This does not seem to be the case, although it is hard to say for sure, since the distances within a galaxy are so much smaller than those between galaxies that such expansion would be very hard to detect.

Although it was Hubble who convinced scientists that the universe is expanding, Albert Einstein nearly reached the same conclusion some years earlier, but, in a tactic he later called his greatest mistake, backed off. When he applied the equations of his relativity theory to the universe, Einstein included the then-standard assumption of a static universe. Under that limitation, the equations said the universe should be shrinking, or collapsing. Since Einstein, like most others, believed in a static universe, he concluded that there must be some yet unknown force by which galaxies repelled one another—a force that precisely offset the collapsing. It was, in a sense, a "fudge factor" that made the equations fit his preconception. Once it became clear that the universe was expanding, Einstein realized his mistake. Had he not imposed the fudge factor, he could simply have left his equations alone and they would have told him that the universe was expanding.

As mentioned earlier, the 1965 discovery of the 3-degree heat of the universe clinched the Big Bang theory. It also won the Nobel Prize for Penzias and Wilson. Although this discovery has often been described as the cooled-off heat of the Big Bang, this is not quite right. The heat is, in fact, the original energy of the Big Bang, but that energy has expanded along with the matter in such a way that it has, in a sense, been diluted by the vast distances now occupied by the universe.

Since the confirmation of the Big Bang theory, debate has focused on two questions. One is whether or not the universe will continue expanding forever. The crucial factor in attempting to answer this question is the amount of mass in the universe. If there is enough mass, its combined gravity should gradually slow the expansion, just as Earth's gravity slows the speed of a ball thrown upward. If this is the case, the far-flung galaxies should, like the ball, eventually slow to a stop and fall back together in a cataclysmic implosion that, theoretically, would compress the entire universe back into a very small point. If this happens, it will be billions of years in the future, since the universe should take as long to collapse as it did to expand. If there is not enough mass to exert the gravitational pull needed to compress the universe, it will continue expanding forever. Depending on various theoretical assumptions, astronomers differ on whether the universe is "open" and will always expand or is "closed" and will eventually collapse.

The second great problem in cosmology today is vastly more complex. It deals with the detailed events of the first moments of the Big Bang. Imagining what the situation was like at the start of the explosion is very nearly beyond human capacity. All the "stuff" that today exists as matter and energy—all the trillions of galaxies with their trillions of stars and planets—were compressed at the very beginning into an infinitesmal speck. The temperatures and pressures that would have existed then were so great that neither energy nor matter could have existed in their present forms. All that existed was a single something (nobody knows what) in which all that exists today was unified. All the various atoms and molecules, all the electrons and protons, all the forces of gravity and electromagnetism, and more had not yet assumed their separate identities, and were still unified.

Once the explosion began, conditions changed rapidly, and the various forces and particles "condensed" or "crystallized" (the words are only metaphors) into different forms. According to a scenario proposed by Sheldon Glashow, an American physicist who won a Nobel prize for his efforts to explain these matters, the most important events occurred within the first few millionths of a second. As the infinitesimal cosmos grew in size, pressures and temperatures diminished, and the various forces and particles came into separate existences. About the time the universe was the size of a bowling ball—only a billionth of a billionth of a billionth of a billionth of a second after the start—electrons and quarks "crystallized." Later, by one ten-billionth of a second, all the known particles and forces (which are carried by other kinds of particles) had come into existence. After about two minutes, the temperature would have fallen to about a billion degrees, and these particles could begin to combine. Neutrons combined with protons to form units that would be the nuclei of hydrogen and helium atoms. Within a few hours after the start of the Big Bang, the temperature would have dropped to only 10 million degrees, too cool to weld any more particles into atomic nuclei.

For the next few thousand years the nature of matter did not change much, although, of course, the infant universe kept expanding rapidly, as a cosmic soup of subatomic particles. Eventually conditions became cool enough that the nuclei generated in the Big Bang began to capture electrons. By about a million years, the temperature of the universe had fallen to around 4000 degrees, and virtually all of the neutrons and protons had captured all the electrons they could hold—leading to a universe that consisted of about 80 percent hydrogen and 20 percent helium. All the other elements would be formed much later inside stars, which had not yet formed.

According to this scenario the early universe was a rather homogenous place—a cosmic soup of evenly distributed particles. The universe today is not homogenous. Matter is organized not only into planets and stars but also into galaxies, and even galaxies usually exist in vast clusters. Somehow the stuff of the primordial universe was perturbed in such a way that it formed clumps. The stronger gravity of the clumps would have pulled in free particles and grown still larger. Eventually—perhaps not until a billion years after the start of the Big Bang—the centers of these accreting clouds of hydrogen and helium would be compressed enough to ignite the same kind of reaction that powers a hydrogen bomb. When this happened, the clumps burst into stars that burned for a few million or billion years and then died out. Since the origin of the universe, many generations of stars have lived and died; new stars and new galaxies are still being formed.

It is within the stars that the simple atoms of hydrogen and helium were welded into bigger atoms to form all of the other chemical elements. The atoms that make up human beings, and practically everything else on Earth, were forged inside stars that lived long ago. These ancient stars burned out, bequeathing a wealth of new, heavier elements that would play new roles in the evolving universe.

Today's universe is still exploding. We are, in a very real sense, still inside the Big Bang. The energy of the early heat still permeates space—the 3-degree background radiation. The only difference is that the heat is now spread out over so much larger a volume—the whole universe—that it is, in effect, diluted.

There are signs that the Big Bang is slowing down, that the universe has not been expanding at a steady rate. The reason it is slowing down is that gravity from each object in the universe tugs at every other object. The effect is precisely the same as the one that slows a ball tossed upward. The ball steadily slows down, stops, and reverses, collapsing, as it were, back toward the Earth. One of the big questions about the universe is whether it, like the ball, will stop and collapse. If there is enough mass to cause the universe to collapse, then we live in what physicists call a closed universe, and the collapse should happen billions of years from now. If there is not enough mass, we live in an open universe and all the galaxies in the universe may eventually move so far apart that the diminishing pull of gravity with distance becomes too weak to stop the expansion.

See ATOM; SOLAR SYSTEM, ORIGIN OF; CLOSED UNIVERSE; HUBBLE; EINSTEIN.

BIOGENESIS—Animals always have parents. Stated that way, the idea seems trivial, even childish to our twentieth-century minds. Just two centuries ago, however, that simple formulation came as a shock, a refutation of popular beliefs that had lasted for thousands of years. Of course at that time it was couched in the respectably pompous Latin of scholars: *Omne vivum ex vivo:* Everything that lives came from something else alive. This is known as the law of biogenesis, and it was not an idea that came overnight. In fact, it took most of a century of dogged experimentation by three scientists to lay low the ancient belief in spontaneous generation. According to that venerable doctrine, maggots would form in rotting meat, earthworms would pop into existence in a pile of manure, and even mice would spontaneously develop in stacks of grain. That, after all, was common sense to everyone in the eighteenth century—everyone, that is, except two skeptical Italians, Francisco Redi and Lazzaro Spallanzani, and one skeptical Frenchman, Louis Pasteur.

In a way it was odd that the idea of spontaneous generation flourished as it did in Christian Europe. After all, it meant that God had not created everything back during those six epochal days of Genesis, but that he was still at it, preferring dung heaps and garbage piles in which to display his handiwork. Still, it was believed.

Redi delivered the first blow in the 1600s by conducting a controlled experiment. He took two jars and put some meat in each. Over one jar Redi placed a gauze veil. Then he watched. Flies soon gathered and could reach the meat in the open jar but not the meat protected by the gauze. A few days later, maggots—the larval forms of flies—appeared in the uncovered meat. The covered jar had no maggots. Redi concluded that flies were laying eggs in the meat they could get at, and that from these—but not from the gauze-protected meat—came new flies.

That experiment pretty well answered the question for animals as big as flies. But the theory of spontaneous generation still hung on with respect to the littlest animals, the microbes that Anton van Leeuwenhoek was discovering up in Holland. Even the gauze covering did not stop them from appearing in the meat.

Spallanzani, working in the 1700s, borrowed Redi's experimental strategy and used it to prove exactly the same point for microscopic organisms such as the bacteria that van Leeuwenhoek had discovered. Gauze wouldn't seal the container, so Spallanzani found a way of sealing the glass to keep out even the dried-up bacterial spores that float in air. (When these get wet, they resume ordinary life processes.) He also boiled the glass containers of water and nutrients such as bean seeds long enough to kill any organisms that had been trapped at the time the glass was sealed. Then he let the containers sit for weeks before examining their contents under a microscope.

In the sealed containers, there was no sign of life. In similar containers treated in exactly the same way but merely stoppered with corks he found teeming bacteria. Spallanzani had proved that even the simplest germs do not arise spontaneously. He had also, if unwittingly, laid the foundation for the canned food industry, which seals its nutrients in glass or metal containers and boils them to kill trapped organisms. If there are no live bacteria in the containers, the food so preserved cannot spoil.

Spallanzani's research was done in the mid- and late-1700s, but nobody at that time caught on to the role of bacteria in causing such everyday phenomena as the fermentation of wine or beer or the spoiling of milk. Not until well into the 1800s did Pasteur come along with his famous experiments, much like those of Spallanzani, to prove that various kinds of bacteria were responsible for these forms of decomposition. Pasteur was, of course, the inventor of the process we call pasteurization. It amounts simply to heating a product—be it milk or beer, or anything else bacteria might like to eat—to a temperature sufficient to kill all or most of the bacteria present in the product. In some cases, such as with milk, however, vigorous boiling would not only kill all the bacteria but destroy the product itself. Pasteurization in such cases involves heating only enough to kill or cripple most of the more troublesome bacteria. This retards decay but does not prevent it entirely.

See REDI; SPALLANZANI; PASTEUR; LEEUWENHOEK.

BIOLOGICAL CLOCK—There are several kinds of biological clocks. Tulip bulbs have one that tells them when to start pushing leaves up to the surface. Women have one that tells them when to ovulate and carry out other events in the menstrual cycle. Virtually all species have clocks that regulate their metabolism in a 24-hour, day-night cycle called the circadian (meaning "approximately daily") rhythm. Human beings have many circadian rhythms, with various hormones and other substances ebbing and flowing throughout the body, depending on the time of day. Some of the clock mechanisms are known; most are great mysteries. A few clocklike rhythms are learned, such as the hunger that develops at mealtime. Most, however, seem to be genetic, but may need an environmental effect to trigger them. For example, some animals kept in light from the time of their birth lack certain circadian rhythms, but if put in the dark for an hour or so, immediately begin cycling through the rhythm they lacked. There is evidence that some animals regulate their clocks by the position of the sun and moon. Fiddler crabs, for example, have an activity rate that varies according to the tides along the beaches where they live. Move them a thousand miles inland to a laboratory, and they shift their activity cycles to correspond to what *would* be a tidal cycle based on the position of the sun and moon if there were an ocean at the new location.

BIOMASS—The weight of all the living things in a given area is the area's total biomass. It is a measure of the abundance of life that a habitat is able to support. Biomass can also be calculated for separate species. In East Africa's Serengeti plains, for example, the biomass of all the animals living above ground—elephants, millions of wildebeeste, lions, and so forth—is less than the biomass of the termites munching the roots of grass below ground.

BIOME—This is the ecologist's word for the largest geographical area dominated by a given form of vegetation. There are grassland biomes, desert biomes, coniferous forest biomes, and the like. A biome may cover many thousands of square miles and include several kinds of plant and animal communities.

BIONICS—Any machine that is made to emulate a behavior of a living organism is bionic. The term was coined by the military in 1958 to apply to various devices such as vehicles that move on jointed legs. Sonar, which operates like the echolocation system of bats, also counts as bionic. Whether computers are bionic is a good question.

BIOTA—A collective term for all the living organisms of a given area. The area may be called a biome. See BIOME.

BLACK HOLE—Surely no concept from theoretical astrophysics has captured the popular imagination as rapidly as have black holes. (The only rival might be the notion of time travel stemming from relativity theory.) Black holes, one must remember, have never been detected for sure. Certain mathematical calculations based on what happens to dying stars suggests that they surely exist, but as yet, nobody has been able to prove their existence, although there are many good candidates in the sky.

A black hole is a lump of matter that has been compressed to such a great density that its gravity is strong enough to keep any radiation—whether light, heat, or radio waves—from escaping. Since no light can escape, from

it, the lump remains black to our eyes. The hole part of
the name comes from the effect that such a strong source
of gravity has on the space around it. Einstein's general
theory of relativity says that gravity warps space. The
path of any object or line of sight that would otherwise be
straight is made to curve in the vicinity of a strong gravitational field. A black hole's gravity is so strong that the
space near it doesn't just warp, it curves right into the
center of the black hole. The curvature of space is a diffi-
cult concept to grasp (it is explained more fully under the
entry on relativity), but three-dimensional models of the
curved space near a black hole (made to represent four-
dimensional space-time—the three dimensions of space
plus one of time) look like a funnel with the black hole
sitting at the bottom—a hole in space-time into which
any object that comes near will fall, pulled by the im-
mense gravity of the lump of matter that comprises the
hole.

Black holes are thought to form when very large stars
die. When any star is young, the heat and radiation from
the thermonuclear fires that make it shine produce enough
outward pressure to keep the star "inflated." Eventually,
however, the nuclear fuel is used up, the outward pressure diminishes, and the weight of the remaining matter
causes the star to collapse from all sides toward the center. The smallest stars simply shrivel uneventfully into
what is called a white dwarf, a hot sphere that may be
smaller than the Earth but which has no nuclear fires.
Larger stars experience a much more cataclysmic demise.
They explode briefly into a supernova, blasting away
much of their matter. If the matter that remains has more
than 1.4 times the mass of our sun, the remnants of the
supernova will collapse toward its own center to become a
black hole. In these cases, the mass is so great that once
the collapse starts, it never stops. It is against all common
sense, but the equations that predict black holes (based on
relativity theory) show that the matter continues collapsing toward an infinitely small point at the center of the
hole, the so-called singularity.

Unfortunately, it is impossible to know what happens, in the final formation of the hole because once the matter shrinks below a certain size (which varies with the original mass), nothing further can be seen. The boundary of this "certain size" is an imaginary sphere called an event horizon. Within the sphere, or event horizon, the matter continues to shrink, but we can never detect what is happening because the event horizon marks the boundary within which gravity is strong enough to keep in all light and radio waves. The matter continues shrinking toward the singularity, but the radius marking the event horizon stays the same.

It seems incomprehensible that matter can collapse to this degree. Some of the collapse, however, may be easier to appreciate when it is clear that most of the volume occupied by any kind of matter is empty space. The particles that make up atoms are farther apart in scale than the planets of our solar system. With a strong enough force, you can squeeze all this empty space out of an atom, shrinking the Empire State Building, for example, to something the size of a sewing needle. It would still weigh as much, though, because all of its mass would still be there. All that has been removed is space. But even this would not be enough compression to create a black hole. You would also have to squeeze out the space between the quarks that make up protons and neutrons. You might also have to squeeze even more than this to crush quarks together, but nobody knows for sure.

If the Earth were squeezed to this degree, and compressed to the size of a golf ball, it would become a black hole. All of its trillions and trillions of tons of matter would still be there, but all of the gravitational force exerted by that matter would be concentrated less than an inch from the center of the mass. The "golf ball" would mark the event horizon within which the Earth's mass continued collapsing toward the singularity.

This last point is important in understanding black holes. The total amount of gravitational pull exerted by a black-hole Earth would be exactly the same as the Earth now exerts. The big difference would be in the distance from the center of the mass to the surface. If you moved from the center of this black hole to a point as far away as the real Earth's surface is today (roughly 4,000 miles), the gravity would be exactly what it is on the real Earth. If you were in a space ship, the pull would tend to drag the space ship in. If you had enough power, you could escape unless you grazed the "event horizon" barely an inch from the center of the black hole. Then the pull would be strong enough to let nothing escape.

Some physicists believe there are countless billions of black holes in space, from giant ones (that is, black holes with spherical event horizons of very large radius) that may reside at the centers of galaxies to microscopic ones that may be drifting about aimlessly in open space. Some physicists even suggest that Earth may have collided with one of these microscopic black holes in 1908, causing the mysterious explosion in Siberia known as the Tunguska event. If so, the black hole probably passed straight through the Earth and flew out the other side, having sucked in as much Earth matter as it could during the moments it took to penetrate.

See GRAVITY; SUPERNOVA; TUNGUSKA EVENT.

BODY BURDEN—As if we didn't have enough burdens, the modern age of toxic chemicals and radiation has given us new ones. Many toxins, like radiation effects, accumulate in the human body, each amounting to a body burden of so many parts per million or, in the case of radiation, so many rads. Every human ever tested has a body burden of DDT. One part per million is roughly the ratio of one jigger of vermouth in a railroad tank car of gin. One rad is not so pleasant to explain. A rad, or "radiation absorbed dose," is a unit of radiation whose effects are retained in the body. About half the people who accumulate a body burden of 450 rads will die.

BOHR, NIELS, 1885–1962, Danish atomic physicist—The first modern concept of how atoms are built—Ernest

Rutherford's model of a nucleus surrounded at a distance by orbiting electrons—contained a fatal flaw. Under the laws of classical physics, the electrons, like old artificial satellites in Earth orbit, should have spiraled down to collide with their nuclei almost immediately. Obviously they didn't. It was Niels Bohr who explained why, and in the process tied the new quantum theory of Albert Einstein and Max Planck into the emerging model of the atom.

Electrons stay in their orbital paths, Bohr found, because to drop gradually to a lower orbit, they would have to lose energy gradually. Quantum theory holds that energy comes only in discrete quantities (quanta) that are lost or gained as whole units. There is no such thing as half a quantum. Electrons, then, are locked in their paths. The electrons are fixed, like trains on parallel tracks without switches to let a train move gradually from one track to another.

Before taking on atomic physics, Bohr conquered soccer. Along with his brother, Bohr was, as a young man, a member of the All-Denmark soccer team and famous all over Scandinavia.

Bohr was born the son of physiology professor at the University of Copenhagen. He showed early talent in science and, upon receiving his Ph.D., he headed for the world's leading center of physics research, which at that time was the Cavendish Laboratory in Cambridge, England. There he studied under J. J. Thomson, who had discovered the electron, and worked with Ernest Rutherford, who became a lifelong friend. So close were the two, in fact, that when Bohr's son was born, he named the boy Ernest, pointedly spelling it the English way instead of the Danish Ernst.

In 1922 the Nobel committee awarded Bohr its physics prize for his modification of his friend's ideas on atomic structure. Bohr, then thirty-seven, was the youngest person to win the prize up to that time. By then he had left the Cavendish and returned home to become head of the Copenhagen Institute for Theoretical Physics. Bohr's award only added to a worldwide fame that drew still more aspiring physicists to the Copenhagen Institute, which rose above the Cavendish in fame as a center of research on the atom.

It was at the Institute that scientists in the free world first learned of experiments going on in Nazi Germany to develop the atom bomb. Worried, Bohr traveled to the United States to talk with Albert Einstein, who had already fled Germany, and other scientists. It was this visit that led Einstein to write his famous letter to President Roosevelt, leading Roosevelt to establish the Manhattan Project to beat the Germans to the new kind of bomb.

Bohr returned to Denmark in time to see his country overrun by the Nazis. Bohr, whose mother was Jewish, felt it best to flee. Along with some 5,000 Danish Jews, Bohr and his family were secretly shipped out of the country aboard fishing boats to Sweden. From there Bohr went on to the United States and to Los Alamos to join the Manhattan Project. Once the atom bomb had been demonstrated, Bohr joined the group of scientists who immediately pleaded for international control of the devastating weapon. After the war he returned to Denmark and became head of his country's atomic energy commission. He worked for the rest of his life in behalf of efforts to prevent further hostile use of the atom bomb.

See ATOMIC STRUCTURE; RUTHERFORD; ELECTRONS; THOMSON, J. J.; QUANTUM THEORY.

BORN, MAX, 1882–1970, German physicist. See EINSTEIN.

BOYLE, ROBERT, 1627–1691, English natural philosopher—One of the founders of modern chemistry, Boyle, is best remembered today for his formulation of what is known as Boyle's law. This states that a volume of gas is inversely proportional to its pressure. In other words, in order to squeeze a volume of gas to half that volume, you must double the pressure on the gas. Quadruple the pressure and you compress the gas to one-fourth its original volume. And so on.

Had he not discovered this law, Boyle would probably
be best remembered as the man who helped an impoverished Isaac Newton get his *magnum opus*, the
Principia, published. In those days scholars paid printers
to publish their books, and Boyle, who had inherited
wealth, gave Newton the money to make his greatest
work known.

Still, Boyle was a pioneer in many other fields of science. He grasped the nature of chemical elements and the
atom theory a century before John Dalton established
these concepts firmly. He also studied the speed of sound,
the phenomena underlying color, and the nature of static
electricity.

Boyle was able to pursue such wide-ranging interests
because he was always financially well off. Although he
was the fourteenth child of the Earl of Cork, his father
was a very wealthy man who saw to it that Boyle got the
best of tutors and schools. After three years at Eton College, England's most prestigious private school, Boyle
was taken out by his parents and, at the age of eleven
sent to tour Europe. In Italy he had the good fortune to
meet Galileo, who convinced the young Englishman, then
but fourteen, that he should devote his life to science.

Back in England, Boyle went to Oxford and fell in with
a group of scientifically minded people whose informal
gatherings eventually received a charter from the king,
creating them as the Royal Society. Today the Royal Society is Britain's most prestigious scientific society.

See DALTON; ATOM THEORY; NEWTON.

BRAHE, TYCHO, 1546–1601, Danish astronomer—
Tycho, as he is usually called, was one of the most eccentric and flamboyant of scientists, and at the same time one
of the most meticulous. He produced the most accurate
measurements of astronomical objects and events ever
achieved before the invention of the telescope. In the
hands of mathematician Johannes Kepler, a student of
Tycho's, the data revealed that the planets move in ellipses around the sun, not the perfect circles posited by Nic-

olas Copernicus almost a century earlier. Kepler's description of planets' orbits, in turn, allowed Isaac Newton to conceive his epochal laws of planetary motion and universal gravitation.

Tycho's extraordinary astronomical measurements emerged from what was by far the most unusual scientific establishment of its time, and perhaps of any other time up to recent years. King Frederick II of Denmark was so impressed with Tycho's promise as an astronomer that he agreed to support the establishment of a vast astronomical observatory commanding the services of many technicians and apprentice astronomers—something the likes of which the world had never seen. It was to fulfill not only a monarch's idle passion but Tycho's uncompromising insistence on the finest quality in everything from the great sextant-like instruments to the food served his staff and guests in the banquet halls.

In 1576 the king gave Tycho a fair-sized island in the Danish Sound and the money to build not only an observatory, but auxiliary laboratories and workshops for constructing instruments. Master craftsmen from around Europe were imported to build the facilities, which also included a windmill, a paper mill, a printing shop, and a bindery. (Tycho not only feared typographical error from outside printers, but wanted the paper and bindings on which his works were published to be of the finest quality.) The living quarters of the observatory even had running water—a novelty in the sixteenth century—thanks to Tycho's invention of a pressure system that pushed the water to upper floors.

Tycho's island establishment flourished for some twenty years, becoming the center of astronomy in northern Europe. But in 1588 King Frederick II died and Tycho fell out of favor with his successor. Lacking funds, Tycho left Denmark and found patronage under the emperor of Prague, where he continued his astronomical work. Kepler became his student and it was to him that Tycho, on his deathbed, willed all his observational data.

See KEPLER; COPERNICUS; NEWTON.

BRENNER, SIDNEY, 1927–   , South Africa-born British biologist. See GENES, HOW THEY WORK.

BROWNIAN MOTION—Tiny particles of a substance dispersed in a fluid may never settle to the bottom, or if they do, may take an extremely long time to do so even if the fluid is allowed to stand quite still. This is because the particles are so small that they are kept in suspension by being constantly bombarded by the molecules of the fluid surrounding them. (The molecules of any fluid are always bouncing about among themselves. The energy of this movement is heat.) The random dance of the suspended particles as they are repeatedly bumped by the fluid molecules was first described in 1827 by Robert Brown, a British botanist, while he was watching pollen grains in water. The motion, subsequently named for Brown, was central to Einstein's epochal 1905 paper in which he finally clinched the proof of atom theory—the idea that matter occurs in discrete particles. The direction in which the Brownian particle moves is determined when a greater number of atoms happen to hit the particle on one side than on the opposite side.

See HEAT; EINSTEIN; ATOM THEORY.

CALORIC—This was a term scientists used in the eighteenth and nineteenth centuries for a mysterious fluid that caused objects to become hot. Caloric had no weight and could not be seen, but seemed to flow from warmer objects into adjacent, cooler objects. Subsequent work on the nature of heat revealed this flow to be a form of energy and not a strange kind of matter.

See HEAT; THERMODYNAMICS; LAVOISIER.

CALORIE—This is a unit of heat energy, not fat. It is the amount of heat required to raise 1 gram of water by 1 degree Celsius at sea level. The calories in food are also units of heat, which the body can use as energy to do work or, if no work is done, which the body will store as fat for future use. Since the physicist's calorie is inconveniently small for measuring the energy content of food, nutritionist's use a larger unit called the Calorie (with a capital C), which is equal to 1,000 calories (with a lower-case c). The Calorie content of a food is determined by literally burning the food and measuring the amount of heat put out.

CARBON-14 DATING—See RADIOCARBON DATING.

CARCINOGEN—In the broad sense, anything that causes cancer can be called a carcinogen, be it radiation, a virus, or a chemical. In the narrower sense, a carcinogen is a type of molecule capable of entering the nucleus of an animal or human cell and binding to the DNA in a way that causes the adjacent gene to be miscopied during the normal process of cell division. The miscopy produces a mutation. If the mutation happens at a certain place on a certain gene, the gene will be converted into one that commands the cell to keep dividing, proliferating indefinitely. The proliferation produces a tumor. Some chemicals loosely called carcinogens do not do this, but do abet the cancer process by speeding the rate of normal cell proliferation. They are called promoters.

CARNOT, SADI, 1796–1832, French physicist. See THERMODYNAMICS.

CATASTROPHISM—Until the middle of the nineteenth century, most people thought the Earth's major geological features—its mountains, valleys, oceans, and so forth—were caused by sudden cataclysmic events, or catastrophes. Noah's flood was the archetypal catastrophe. Then Charles Lyell, a British geologist, came up with an alternative concept that turned out to be correct. It was eventually called uniformitarianism because it asserted that geologic features were caused by gradual events continuing more or less uniformly over very long periods. See LYELL; UNIFORMITARIANISM; PLATE TECTONICS.

CAVENDISH, HENRY, 1731–1810, British physicist— Henry Cavendish was such a perverse recluse—rarely appearing in public or even publishing his results—that it was not until seventy years after his death that people fully appreciated him. Cavendish's notebooks revealed that he had discovered but never published a number of important principles of electricity decades before others made the same discoveries and got the credit. He discovered, for example, that the force between a pair of electrical charges is inverse to the square of the distance between them, a finding repeated much later and eventually named Coulomb's law for the one who published it. In the same sense Cavendish also discovered Ohm's law, which relates electrical current and the resistance to its flow, almost half a century before Ohm. Besides this, Cavendish anticipated or came very close to other great discoveries involving the chemistry and physics of gases and the nature of heat.

The work for which Cavendish alone is remembered— mainly because he did publish his findings—was the discovery that water was not an element (not an indivisible unit of matter), but a compound made of two elements: hydrogen (two parts) and oxygen (one part). Cavendish understood that when hydrogen is burned, or oxidized, the process synthesizes water. He also concluded that air was not an element but a mixture of four parts nitrogen and one part oxygen. (He was only slightly off. Air is also about 2 percent other gases such as carbon dioxide, argon, and so forth.)

Cavendish's reclusiveness is legendary. He almost never appeared in public, and then chiefly at scientific meetings. Even there he said little, earning a reputation for silence rivaling that of Trappist monks. He never married, living instead in London with his father, an administrator of the British Museum. He was, in fact, such a misogynist that he ordered all his female servants to stay out of his sight; he communicated with them only through written notes. There is no indication Cavendish ever formed close attachments to anyone outside his own family. When he was forty, he inherited a millionaire's fortune (the Cavendishes had nobility among their kin), but he and his father continued to live modestly, spending the money on scientific equipment and books. Cavendish willed his fortune to various relatives. In 1871, sixty-one years after the scientist's death, the Cavendish family used a good portion of the money to endow the Cavendish Laboratory at Cambridge University, where their benefactor had studied. The laboratory has become one of the centers of modern research in physics.

CELL (ELECTRICAL)—Though often called a battery, the chemical device that produces electricity is actually a cell. A battery is, properly speaking, a gang of cells. Each cell is a self-contained chemistry set built with one thing in mind—to make certain that atoms shed electrons that can then flow out from one terminal of the cell and back in through the other. This flow of electrons is electricity. Each cell of a battery consists basically of two different substances, each attached to one post, or terminal (usually metals) immersed in a third substance (often a liquid, called an electrolyte, such as the acid in a car battery). One terminal is attached to one of the immersed substances, say zinc, which dissolves in the electrolyte, releasing the electrons that once bound the zinc atoms together. The free electrons then accumulate in the zinc until there are so many that they stop the dissolving action of the electrolyte. Meanwhile, the other terminal is attached to another metal, such as copper, that is immersed in an electrolyte that tends to remove the free copper atoms from the metal. This "uses up" electrons, producing a chronically electron-poor situation in this second metal. Thus, the terminals of a cell are linked to chemical environments in which opposite kinds of reactions are "trying" to take place. One terminal is accumulating surplus electrons while the other has a deficit of electrons. If the two terminals are then linked via a wire, the electrons from the zinc flow across to the copper, relieving the surplus on one side and replenishing the deficit on the other side. A battery cell runs down as the atoms available for reaction in either electrode are depleted or when the electrolyte becomes so saturated with the liberated electrode atoms that the reaction cannot proceed. Recharging binds the loose atoms back onto their electrode of origin.

Almost any combination of dissimilar metals can act as an electrical cell (though with varying degrees of strength)—a fact discovered in 1800 by Alessandro Volta, an Italian physicist who invented the battery and for whom the volt was named.

See ELECTRICITY; ELECTRON; VOLTA; GALVANI.

CELL (LIVING)—Almost every part of every living organism is made up of many specialized cells, a few hundred or thousand in the case of very small animals, a few million or billion in the case of big organisms such as human beings. Most cells are microscopic in size, but some are big. Certain human nerve cells, for example, consist of a thin fiber three feet long. The largest cell in the animal world is an egg yolk. The internal structures it possesses in common with most other cells—a nucleus to hold the DNA and assorted other components for carrying out the DNA's instructions—are dwarfed by the accumulated product of one of its DNA-dictated functions, the synthesis of protein to feed the developing chick embryo.

The smallest animals consist of only one cell with an unusually broad set of abilities and specialized components. Examples include the amoeba and the paramecium, both denizens of ordinary pond water. Other one-celled animals sometimes assemble into clusters and behave as an organized colony, with various cells taking on specialized functions. Then, after living this way a while, the colony will break up into its individual cells, each going its own way. Other animals, such as sponges (the kind that live in the ocean, not the ones from the grocery), appear to be permanent colonies of cells. Often the only types of cells in these permanent colonies that seem to enjoy a taste of the free life are the sex cells, such as the free-swimming

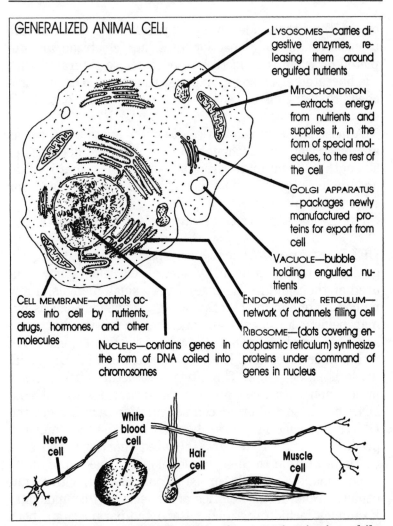

GENERALIZED ANIMAL CELL

LYSOSOMES—carries digestive enzymes, releasing them around engulfed nutrients

MITOCHONDRION—extracts energy from nutrients and supplies it, in the form of special molecules, to the rest of the cell

GOLGI APPARATUS—packages newly manufactured proteins for export from cell

VACUOLE—bubble holding engulfed nutrients

ENDOPLASMIC RETICULUM—network of channels filling cell

RIBOSOME—(dots covering endoplasmic reticulum) synthesize proteins under command of genes in nucleus

CELL MEMBRANE—controls access into cell by nutrients, drugs, hormones, and other molecules

NUCLEUS—contains genes in the form of DNA coiled into chromosomes

Nerve cell

White blood cell

Hair cell

Muscle cell

sperm of many species. Perhaps the most intriguing philosophical issue involving cells is whether multi-celled organisms are true individuals or whether they are instead colonies of many highly specialized individuals. Did evolution create big animals by converting one-celled individuals into permanent colonies? Many animals, including human beings, not only have free-swimming sperm cells, but white cells in their blood, which travel about the circulatory system, gobbling up foreign proteins and waste chemicals as if they were amoebas in a pond.

CELL THEORY—On first reading, it strikes many people as trivial to speak of cell theory as a major achievement of science. What is so earth-shaking about the idea that complex organisms are made up of many smaller units of life called cells? The fact is that without cell theory—which developed over a period of about seventy years from its earliest conception to the modern form that emerged around 1900—biologists would have been unable to understand how organisms grow and develop from conception. It would have been impossible to figure out how heredity works. And there would have been no way to understand most human diseases, cancer in particular.

Robert Hooke, an early English physicist, coined the term "cell" in 1665 after looking at a slice of dried cork through a microscope and seeing rectangular chambers that reminded him of monks' cells in a monastery. Still, he simply thought of cells as voids, and well into the nineteenth century the term merely referred to tiny holes in living tissues. Hooke's cells were, in fact, the empty spaces inside the cell walls of the cork tree's dead and dried bark. The modern concept of cells as units of life did not really begin to emerge until about 175 years after Hooke. In 1839 two German biologists, Theodor Schwann and Matthias Schleiden, were the first to propose that cells are the elementary particles of life, the smallest components of an organism that may be said to be alive.

Until Schwann and Schleiden came along, most scientists thought that the basic structures making up organisms were fibers and vessels that simply grew, in much the same way as crystals grow, some said. The orderliness with which organisms grow—always forming a variety of specialized organs organized in predictable ways—was attributed to a mysterious, non-material property called the vital force. It was a kind of early scientific euphemism for God. Schleiden even went so far as to perceive one of the more fascinating philosophical questions of cell theory: if a cell is a unit of life, then is a multicellular organism really only a colony of cellular individuals? Some plants and animals, in fact, live part of their lives as free-roaming indi-

vidual cells, and only later congregate to form colonies
that have the characteristics of multicellular organisms.
Speculation about this quickly found its way into debates
in political philosophy about the role of the individual person in relation to society as a whole.

One error in Schwann's thinking was his idea that cells
arise *de novo* out of some generalized body fluid. In 1855 a
major advance in cell theory laid this idea to rest. Rudolf
Virchow, a German pathologist and strong proponent of
the cell theory, suggested instead that all cells are formed
by the division of pre-existing cells. It was a most fundamental modification of cell theory, for it opened the door
to an understanding of how organisms develop, growing
from one cell to many.

Over the next few decades, Virchow's ideas encouraged
more detailed study of cell division. Improved microscopes
and better methods of staining cells (so their components
could be seen more easily) led to a steadily improving
knowledge both of the internal architecture of the cell and
the events of cell division. By 1882 Walther Flemming, a
German biologist, produced the first detailed description
of the process of cell division, or mitosis, as it is called.
Flemming reported that the actual splitting of the cell was
merely the last event in the process. Before it happened,
the chromosomes somehow doubled their number and seg-
regated into two equal sets, which migrated to opposite
ends of the parent cell.

By 1900 cell theory had led to a coherent understanding
of how a complex organism arises. It was clear that the
organism begins as a single cell, formed at conception
when a sperm, carrying a set of genes from the father,
penetrates an ovum, or egg, carrying a complementary
set of genes from the mother. Since normal cells contain
two complementary sets of genes, and since each sperm
and egg carries only one of the two sets, their fusion at
conception starts a new life by reconstituting a full gene
set from the two complementary sets. Only after this dou-
ble-gene complement is formed can the cell formed by fusion of the sperm and the egg divide into two, those two

into four, and so on until a complete new organism exists with billions of cells, all having inherited identical copies of the original set of genes bestowed at conception. Thanks to the fully developed cell theory, biologists understood that at various stages in early embryonic development, the new cells formed by repeated divisions sometimes changed their appearance, becoming specialized for their roles as various kinds of tissues.

Despite this specialization, cell theory made it clear that all cells shared the same chemistry and carried out their internal processes in the same ways. In fact, cell theory holds, correctly, that the cells of all living organisms share an underlying set of metabolic processes that sustain life. Modern molecular biology has confirmed this insight by showing that all living cells, whether they be bacteria, cells in plants, or cells in animals, contain virtually identical genes for the enzymes (all proteins) that carry out many basic metabolic processes. These include such housekeeping functions as breaking down complex molecules to release the energy locked up in their chemical bonds, and assembling simpler molecules, such as amino acids, into more complex ones, such as proteins.

A strict interpretation of cell theory leads to the conclusion that viruses are not living organisms. Bacteria, on the other hand, are alive because within their one cell is all the machinery needed to carry out the metabolic processes of life, including reproduction. Viruses, by contrast, have none of the machinery needed to reproduce themselves. They consist essentially of a short length of nucleic acid carrying a few genes, and a protein jacket around those genes. Viruses can reproduce only by invading an ordinary cell, "hijacking" its machinery for reading genes, and using this to synthesize new genes and proteins according to the instructions coded by the viruses' few original genes.

Modern biology has lent even more importance to cell theory by discovering that the membrane that encloses the cell is much more than a mere container. Instead, the cell membrane is studded with specialized molecules that act as receptors, permitting only certain other molecules to enter or leave the cell. Different kinds of cells, say a liver cell or a nerve cell, have different sets of receptors—dictated by those particular genes that are active in the cell—and thus carry out only the chemical reactions appropriate to their specialized function.

CHAIN, ERNST B., 1906–1979, German-born British microbiologist. See FLEMING.

CHARGAFF, ERWIN, 1905– , American molecular biologist. See DOUBLE HELIX.

CHARGE—For all the familiarity with which scientists speak of objects and subatomic particles as having a positive or negative charge, the fact is that nobody has the faintest idea of what "charge" is. It simply exists. We know this because experiments show that there is some kind of property that comes in two forms. Some objects have one form, some have the other, and some have neither. Objects having the same charge (be it negative or positive) repel one another, while objects having different charges attract each other. The principal natural units of charge are carried by electrons and protons. Each electron always has one unit of negative charge and each proton has a positive charge of equal strength. (Neutrons, as the name implies, are neutral, having no charge.) The designation of negative and positive is entirely arbitrary, although universally accepted. The terms were chosen simply because they represent linguistic concepts that behave something like charges. If, for example, you put one proton together with one electron, their charges cancel out, and the resulting particle (a hydrogen atom in this case) is neutral. Any object that has more electrons than protons has a negative charge, and vice versa.

The attraction of opposite charges is one of the forces that keeps electrons in orbit around the nucleus of an atom. Exactly how this works is not entirely clear, but it is thought to result from the oppositely charged particles continually exchanging other particles called virtual photons.

See ATOM; ELECTRON; PROTON; ELECTROMAGNETISM; CELL (electrical).

CHEMICAL BONDING, THE NATURE OF—If the world is made of only ninety-two kinds of atoms representing just ninety-two elements—as it is—how can there be thousands of different types of matter? The answer, of course, is that most of the material stuff of the world around us is not made purely of one kind of atom or another; it is made of groups of atoms that are bound together in various combinations to make molecules.

The various combinations in which atoms may join runs into the millions, to make everything from the ink on this page (which is actually a mixture of many kinds of molecules) to the innards of your brain cells (which consist of very many kinds of molecules). And yet all the incredible variety of the molecular world is produced entirely by two kinds of chemical bonding, both of which are surprisingly simple to understand.

The reason atoms stick together, or as chemists put it, the reason they form chemical bonds, has to do with the behavior of their electrons—the subatomic particles that whirl around the nucleus of every atom. To understand chemical bonding, you must first know a little about the structure of an atom. There is a full entry on atomic structure, but for now it is enough to read what follows.

Hydrogen, the simplest atom, consists of one electron orbiting a nucleus made up of one proton. Protons are nearly 2,000 times heavier than electrons, and have one unit of positive electrical charge. Electrons, though smaller, have exactly one unit of negative electrical charge. Since opposite charges attract, the electron is held near the proton in somewhat the same way that gravity holds the Earth near the sun.

The same mechanism works in all atoms. The main differences among atoms are in the number of protons in the nucleus and the number of electrons orbiting the nucleus.

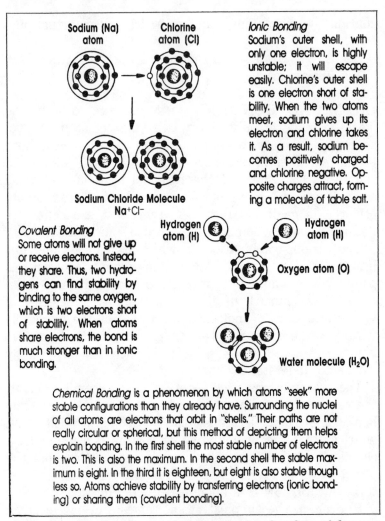

Sodium (Na) atom

Chlorine atom (Cl)

Sodium Chloride Molecule
Na+Cl−

Ionic Bonding
Sodium's outer shell, with only one electron, is highly unstable; it will escape easily. Chlorine's outer shell is one electron short of stability. When the two atoms meet, sodium gives up its electron and chlorine takes it. As a result, sodium becomes positively charged and chlorine negative. Opposite charges attract, forming a molecule of table salt.

Covalent Bonding
Some atoms will not give up or receive electrons. Instead, they share. Thus, two hydrogens can find stability by binding to the same oxygen, which is two electrons short of stability. When atoms share electrons, the bond is much stronger than in ionic bonding.

Hydrogen atom (H)

Hydrogen atom (H)

Oxygen atom (O)

Water molecule (H₂O)

Chemical Bonding is a phenomenon by which atoms "seek" more stable configurations than they already have. Surrounding the nuclei of all atoms are electrons that orbit in "shells." Their paths are not really circular or spherical, but this method of depicting them helps explain bonding. In the first shell the most stable number of electrons is two. This is also the maximum. In the second shell the stable maximum is eight. In the third it is eighteen, but eight is also stable though less so. Atoms achieve stability by transferring electrons (ionic bonding) or sharing them (covalent bonding).

As long as the numbers of these two kinds of particles are equal, their opposite charges cancel out, and the atom is electrically neutral. Oxygen, for example, has a nucleus of eight protons (and eight neutrons, but they don't affect chemical bonding since they are neutral in charge) balanced by eight orbiting electrons.

One more technical point about atomic structure is necessary. The electrons in any atom are not randomly distributed. They exist and move only within definite cloudlike shells enclosing the nucleus. The first shell of any atom, counting from the inside out, can hold only two electrons before it is "full." The next shell can hold only eight. The third can hold eighteen, the fourth, thirty-two.

If the outermost shell of an atom is full of electrons, the atom is chemically stable and will not form chemical combinations, or bonds, with any other atom. The only atoms in which this is the case are those of the so-called noble gases—helium, neon, argon, krypton, xenon, and radon. (They are "noble" because, like nobility, they remain aloof, refusing to join with other atoms.) All other atoms are commoners, with incomplete outer shells, but they try to behave like nobility. They do so by forming bonds with still other atoms that can either donate enough electrons to complete their outer shells or with which they can share electrons in such a way that their shells are completed. These two possibilities—sharing electrons or donating and receiving electrons—are the two forms of chemical bonding by which all molecules are made. Chemists call electron sharing "covalent bonding," and electron donating-receiving "ionic bonding." Covalent bonds, the stronger of the two, are what make two hydrogen atoms stick to one oxygen atom to produce a molecule of H_2O, or water. Ionic bonds—much weaker than covalent bonds—are what make one sodium atom stick to one chlorine atom to create a molecule of Na^+Cl^- (Na abbreviates natrum, the old name for sodium, and the plus and minus signs indicate the charge), or sodium chloride, or table salt.

Let us first consider covalent bonding. Hydrogen, as we have seen, has only one electron. And since the first shell of any atom needs two electrons to be complete, a hydrogen atom will bind to any other atom from which it can get another electron. However, the other atom also has some say in the matter. It, too, is "seeking" to make its outer shell complete. As it happens, many a hydrogen atom has found stability by sharing its electron with another hydrogen atom. Each contributes one electron to the partnership, and each completes its outer (and only) shell with two electrons. The molecule formed is H_2, hydrogen gas.

The behavior of electrons is rather more complex than this book can go into (the entry on quantum mechanics may help), but to understand covalent bonding it is sufficient to visualize the electrons shared by two hydrogen atoms as orbiting around the nuclei of both atoms and, in so doing, keeping them tightly bound together.

Other binding opportunities are also open to hydrogen atoms. Consider the oxygen atom with its eight electrons. Two are in the first shell, filling it. And the other six have no place to go except to the second shell, which has a capacity of eight. Thus, the oxygen atom has two vacancies in its outer shell, and will bind to any atom that has two electrons to share. Or, as commonly happens, it will bind with two other atoms that each have one electron to share. Hydrogen fills the bill, and the result is H_2O. From each hydrogen, the oxygen atom gets the use of one electron, and in return allows one of its electrons to be shared with the hydrogen.

The other way in which atoms can bind is by ionic bonding, so called because it occurs after each atom is turned into an ion. An ion is simply an atom that has lost the balance between its number of protons and its number of electrons. Since these particles have equal but opposite charges, an atom that has an equal number of them carries no net charge. If an atom loses one of its electrons, however, there will be one unbalanced proton in the atom's nucleus, and the atom will have a net charge of $+1$. Conversely, if an atom should gain an extra electron, it will acquire a net charge of -1. A loss of two electrons leaves a charge of $+2$. (Atoms often change their number of electrons but never their number of protons, except when the nuclei of radioactive atoms disintegrate.)

Common table salt—sodium chloride—offers a good example of an ionic bond. The sodium atom has eleven protons and, in its neutral state, eleven electrons. Again, to see how one atom may bind to another, it is necessary to examine the occupancy rate in the various electron shells. Sodium has two electrons that fill its first shell, eight to fill the second shell, and just one electron in its third shell.

Since the third shell has a capacity of eighteen electrons, the lone occupant can wander away very easily. (The fewer electrons there are in a shell, the more readily the electrons will simply fly away.) Once sodium loses its outermost electron, its remaining shells form a very stable unit, but the atom as a whole is left with a net charge of +1 (because there is now one more proton than electron). The sodium atom is now a sodium ion.

Chlorine happens to have the opposite problem. It normally has seventeen electrons arrayed as follows: two in the first shell, eight in the second shell, and seven in the third shell. This is well short of the eighteen electrons it takes to fill the third shell, but only one short of the eight that will produce an intermediate degree of stability. (Shells, as it happens, have subshells, which have their own stable numbers of electrons.) Thus, chlorine has an affinity for loose electrons. If it can find one somewhere, it will snap it up to fill the third shell's subshell. Chlorine atoms that happen to be around sodium atoms will find a ready supply of loose electrons (since sodium easily loses its lone outer electron). The result will be a chlorine atom in which electrons outnumber protons by one, producing a net charge of −1. The chlorine atom is now a chlorine ion, which chemists call chloride.

Since opposite charges attract, the sodium ion (+1) will be drawn to the chloride ion (−1), and they will stick together because of ionic bonding. The result is as if the sodium atom had donated one electron to the chlorine atom.

Although the examples given here involve few atoms, the same mechanisms operate in all molecules.

See ATOM THEORY; ATOMIC STRUCTURE; DALTON (who helped develop these ideas).

CHROMOSOMES—These are the microscopic structures that carry the genes in every living cell except those of bacteria. Chromosomes are made of DNA, a long, threadlike molecule that is, in fact, a chain, or polymer, of smaller units called nucleotides. A gene consists of a series of nucleotides—from a few hundred to several thousand. Many genes, linked end to end like boxcars, form a chromosome.

In the nucleus of every human cell are forty-six chromosomes, arranged in twenty-three pairs, one member of each pair having come from each parent. Ordinarily the chromosomes rest within the nucleus like so much tangled spaghetti. If they were all taken out and laid end to end, they would cover an astonishing three feet in length. The more common image of chromosomes, however, is the shriveled, compact form they take in preparation for cell division. What happens here is that the long threads of DNA (already in the form of a double helix) contract into coils. These coils keep contracting and form coils of coils. The whole effect is rather like what happens when you keep twisting a rope and it begins knotting itself up. Along with DNA there is a protein core to the chromosome that acts as a kind of structural support.

See DNA HOW ITS ROLE WAS DISCOVERED; DOUBLE HELIX; GENES, HOW THEY WORK.

CLASSIFICATION of species—For the system by which all living organisms are classified into species, genera, families, and so on, see TAXONOMY.

CLAUSIUS, RUDOLF, 1822–1888, Prussian physicist. See THERMODYNAMICS.

CLONE—Botanists were using this term back in the 1900s, long before it burst on the scene a few years ago as part of a hoax suggesting that human beings could be duplicated through some biological wizardry. Actually a clone is simply any organism, plant or animal, that has come into existence by non-sexual reproduction. Inheriting all its genes from one parent, it will be genetically identical to that parent. A cutting from Aunt Millie's begonia can be rooted to produce a new individual, which will be a clone of the parent plant. Many plants can be cloned in this way. Bacteria and one-celled animals clone themselves naturally. Attempts in the lab to clone higher animals have enjoyed only limited success. Biologists, for example, have taken cells from tadpoles, extracted the nuclei (where the genes are), and put them into fertilized frog eggs after removing the fertilized nuclei from these eggs. The transplanted tadpole nuclei behave as if they were fertilized nuclei and begin embryonic development. The result is an adult frog that is genetically identical to the tadpole that donated the nucleus. Similar feats have been tried on mammals, but with no success as yet. Identical twins are, however, clones of one another.

CLOSED UNIVERSE—The universe is either open or closed. Nobody knows which, but the issue is one of the most fundamental in all of cosmology. The question is whether or not there is enough mass in the universe to have a combined gravitational pull that can slow the current expansion of the universe to a stop and cause all the galaxies to begin falling back toward the center. If there is too little mass to do this, we live in a open universe that will go on expanding forever. If there is enough mass to stop the expansion, we live in a closed universe that will some day (billions of years hence) slam all of its parts back into a cataclysmic inferno, destroying not only stars and planets, but causing atoms and even subatomic particles to disintegrate. Cosmologists differ on how to calculate the mass of the universe, and have not settled this question.

See BIG BANG; UNIVERSE, STRUCTURE OF.

CODON—This is the geneticists' term for the segment of a gene that carries the code for one amino acid. Each codon specifies which of twenty amino acids is to be added next to the chain of amino acids that will become a protein molecule. A codon consists of three bases, molecules which serve as "letters" in a three-letter word of the genetic code. See GENES, HOW THEY WORK.

COMET—No expression describes the typical comet better than "dirty snowball." Basically, that's what comets are—chunks of ice and frozen gases mixed with rocky debris. The snowballs range in size from 100 yards in diameter to fifty or sixty miles. Some comets orbit the sun, while others may be interstellar wanderers that happen to get pulled toward the sun when they drift into our solar system. Depending on the angle at which they come in, these comets may be captured and held in sharply elliptical orbits, or they may be flung back out of the solar system, never to return.

Most of the time comets are cold and dark. But some put on a spectacular display as they approach the sun. Its light illuminates them, its heat evaporates some of their ice, and the solar wind—a draft of electrically charged particles—blows the comets' evaporated gases into a long, glowing tail. Many comets have no tails; apparently, they are not made of the "right stuff." But when tails form, they may stretch across millions of miles. The record seems to be a comet tail that was more than twice as long as the distance from the sun to the Earth. When a comet leaves the sun, its tail precedes it, since the solar wind blows faster than the comet's head moves.

Until the seventeenth century, each comet was thought to be a unique event, never to be repeated. But Edmund Halley, a British astronomer, predicted that the comet of 1680 would return in 1758. Halley died before his prediction could come true, but it did. Halley's comet is not perfectly regular, but returns approximately every seventy-six years. Since Halley's time, a search has turned up records of his comet's appearance in some culture or other for all but one of its appearances since 240 B.C.

CONDITIONED RESPONSE—This is a tricky subject, fraught with political and philosophical overtones. The original concept involved Pavlov's discovery that if you rang a bell just before giving food to a dog, it would soon learn to associate the bell with the imminent arrival of food, and would begin salivating. Normally, salivation

would start only when the dog saw the food. The dog had
been conditioned to salivate at the sound of the bell.
Clearly an animal, and by extension a human, could be
made to behave in a predictable way by manipulating its
environment, without the need for any coercion. Could political despots employ similar means to make people behave according to their dictates? Such were the
speculations that inspired such books as *Brave New World*
and *1984*.

Some psychologists, eager not to contribute to des-
potism, insisted that Pavlov's discovery applied only to reflexes—behaviors that operate without conscious control.
More complex human behaviors, they insisted, were
learned through entirely different mechanisms. Though
this is probably true to some extent, many psychologists
no longer see such clear-cut distinctions between classi-
cally recognized reflexes and other behavior patterns.
Conditioned responses may play a role in all forms of
learning and behavior.

See PAVLOV.

CONSERVATION OF ENERGY—"Energy can neither
be created nor destroyed. It can only be changed from one
form to another." Like the related law on the conservation
of matter, this principle was articulated by several people
in several ways, from the seventeenth through the nineteenth centuries. Basically, it embraces the fact that, for
example, mechanical energy, such as that which moves
muscles or machines, comes from chemical energy stored
in food or fuel, and that this chemical energy comes originally from solar energy. Although conversions from one
form to another are never 100 percent efficient, the energy "lost" is not destroyed; it is simply converted into a
form, usually heat, that dissipates and cannot easily be
recovered. This unavoidable loss of energy is entropy, the
total amount of which always increases in the universe.

Although the principle of energy conservation is true in
the everyday world, Einstein showed that it must be mod-

ified because energy *can* be converted into matter, and vice versa.

See KELVIN; LEIBNIZ; HELMHOLTZ; ENTROPY; EINSTEIN.

CONSERVATION OF MATTER—"Matter can neither be created nor destroyed. It can only be changed from one form to another." Thus went one of science's classic dictums, perceived by many over the last two or three centuries, but most clearly by two eighteenth-century chemists, Antoine Lavoisier of France and John Dalton of England. Essentially, the idea is that matter consists of units that can be taken apart and reassembled into different combinations. There are circumstances in which it may look as if one kind of matter disappears—the gasoline from the car's tank, or the log from the fireplace—but it is in fact only changing its appearance because it is changing the partners with which it is combined. The units of which matter consists are, of course, atoms or molecules, depending on the situation. In 1905, Einstein's special theory of relativity led to the suggestion that matter could, after all, be changed into something else—energy. And vice versa. That's exactly what nuclear power plants do every day. It is also what every star does.

See LAVOISIER; DALTON; EINSTEIN.

CONTINENTAL DRIFT—The movement of continents over the Earth's surface is one of the effects, along with sea-floor spreading, of the phenomenon known as plate tectonics. One result of continental drift is that North America is moving away from Europe at the rate of about one to two centimeters a year. The theory of plate tectonics was first articulated early in this century by Alfred Wegener, a German geologist, but it was widely dismissed. Today, with abundant evidence, it is accepted as a brilliant insight.

See PLATE TECTONICS; WEGENER.

CONTROL GROUP—Subjects in an experiment (be they chemicals or people) whose reactions are compared to those seen in the experimental group. Experiments without controls yield untrustworthy results.

See EXPERIMENT; PART 1: WHAT IS SCIENCE?

COPERNICANISM—An old name for the theory that the Earth is constantly rotating on its axis and, with the other planets, orbiting the sun. Galileo proved that the Copernican theory was true, overthrowing the older Ptolemaic theory of an Earth-centered universe.

See COPERNICUS; KEPLER; GALILEO.

COPERNICUS, NICOLAUS, 1473–1543, Polish astronomer—Of all the cultural effects of science, few rank with that perpetrated by Nicolaus (also spelled Nicholas) Copernicus. Until his time, the Earth sat serenely immobile as the ordained center of the universe. For at least 1800 years the Western world had accepted the Earth-centered cosmology taught by Aristotle and mathematically explained (awkwardly, to be sure) by Ptolemy, the second-century A.D. Greek astronomer from Alexandria, according to which the sun, moon, planets, and stars all moved about the Earth.

Copernicus proposed instead that the sun was at the center of the system and that the Earth orbited it along with the other planets. Actually, certain Greeks had said the same thing in ancient times, but their views had largely been forgotten, swamped by the enormous influence of Aristotle. Copernicus correctly suggested that the Earth completed one orbit every year, and also that the Earth revolved completely about its axis once a day. Copernicus didn't have a telescope and couldn't make the detailed observations that would prove his theory (Galileo would do that some seventy years hence), but it is to Copernicus that scientists generally credit the shift in philosophical perspective that followed when Earth suddenly seemed to be just one of many planets—a shift known as the Copernican revolution. Copernicus gets the credit for this even though he was wrong on one important point: He stuck to the old Greek idea of planets moving in perfectly circular orbits. Almost a century later, Johannes Kepler would prove that the orbits were actually ellipses.

Although the philosophical shift in the view of the universe had begun, its backlash was not to come until the next century, when the Holy Inquisition attacked Galileo. By contrast, when Copernicus, who was a canon of the Church, told the Pope of his theory, the pontiff approved. Ironically, churchly condemnation of his theory came from Martin Luther, whose strict theological interpretations on other issues would soon also run counter to the Pope's.

Copernicus was the son of a Polish merchant of some social standing. He studied both in Poland and in Italy, eventually emerging as a master of virtually all knowledge of the time, not only in astronomy and mathematics but also in medicine and theology. For a time he worked as a doctor, but for most of his professional life he made use of his degree in canon law by serving as canon of the cathedral in Frauenburg. This was a modest sinecure which allowed Copernicus to pursue his chief interest, astronomy.

Although Copernicus began to crystalize his theories by about 1514, he was reluctant to publish them. He did, however, circulate papers privately among friends. Not until 1540, however, did his friends persuade him to publish his theories. But before the first edition was printed, supervision of the work fell into the hands of one Andreas Osiander, who disliked its refutation of the hallowed, Earth-centered cosmology. On his own authority Osiander inserted a preface claiming that the sun-centered theory was only a mathematical convenience that made the calculations come out right, not actually an assertion that the Earth moved around the sun. Copernicus did not see the finished book until a copy was brought to him just hours before he died.

See GALILEO; KEPLER.

COSMIC RAYS—These are particles from outer space—chiefly protons but also electrons and other subatomic components—that continually bombard the Earth's upper atmosphere at speeds close to that of light. Often these primary cosmic rays hit atoms of the air (usually nitrogen), knocking loose a shower of other particles called secondary cosmic rays. (Sometimes when cosmic rays hit a nitrogen atom, they substitute a neutron for a proton in the nucleus of the atom, converting the nitrogen into an unstable carbon atom that will some day decay back into nitrogen. This radioactive carbon, called carbon-14, is useful in determining the age of objects in archeological sites.) Some primary cosmic rays may penetrate all the way to the ground, but more often it is the secondary particles—such as pi mesons (or pions) and mu mesons (or muons)—that reach the ground.

Nobody knows for sure where cosmic rays come from, but it is believed that supernovas, the explosive, dying phases of smaller stars, spew out great quantities of such particles. Cosmic rays that reach the ground may cause genetic mutations by hitting the atoms within a gene and changing one "letter" of the genetic code to another "letter." Most often, such mutations will be harmful, at least to the cell that the rays hit, and if the mutation is in the sex cells that produce sperm or ova, it may affect entire subsequent generations of the organism. Evolution may be driven by cosmic rays.

See ATOMIC PARTICLES; RADIOCARBON DATING; SUPERNOVA; MUTATION; EVOLUTION.

COULOMB, CHARLES A., 1736–1806, French physicist. See CAVENDISH.

COVALENT BONDING—See CHEMICAL BONDING.

CRICK, FRANCIS, 1916– , British biophysicist—Though he is best known as the second person on the 1953 paper with James Watson, an American molecular biologist, that announced the double-helix structure of DNA, Crick was responsible in his own right for a major part of deciphering the genetic code.

Once its double-helix structure was clear, biologists could see immediately how the long, threadlike molecule of DNA could duplicate itself in preparation for cell division. But it remained a mystery how the molecule carried its genetic messages. It was Crick, after breaking up his partnership with Watson, who figured this out.

See DOUBLE HELIX; DNA; GENES, HOW THEY WORK; WATSON.

CRITICAL MASS—This is the concentration of radioactive material needed so that the decay of one atom—a process in which an atom emits a component particle, along with some energy, and changes into a lighter, different atom—triggers the decay of only one other atom, thus sustaining a chain reaction that continues at a steady rate. When a nuclear power plant "goes critical," it allows its radioactive fuel to achieve a steady chain reaction, producing an even release of energy. In an atomic bomb, by contrast the mass is supercritical, and the decay of one atom triggers the decay of many others, producing a rapidly growing release of energy.

In nuclear fission, a uranium atom hit by a neutron becomes unstable, breaks into smaller pieces, and emits several more neutrons. If all of these neutrons are allowed to reach other uranium atoms, the chain reaction will quickly mushroom, becoming supercritical. To filter out excess neutrons, nuclear power plants insert various materials between the rods containing the uranium atoms. These materials absorb neutrons, thus "moderating" the chain reaction and keeping the mass just at the critical level.

See RUTHERFORD (who produced the world's first nuclear reaction in the lab in 1919), and FERMI (who led the first effort to create a controlled, sustained chain reaction in 1942).

CUVIER, GEORGES, 1769–1832, French zoologist—Georges Cuvier was the first scientist to take fossils seriously, examining them for evidence of ancient life forms. He thus established paleontology as a science and laid the groundwork for Darwin's theory of evolution a generation later. Though he did not live to learn of Darwin's work, Cuvier did dispute the evolutionary hypothesis of his colleague Jean-Baptiste Lamarck who argued in 1809 that evolution was a fact (although his idea of how it happened was wrong). Cuvier held that a living organism was such a finely tuned combination of structures and functions that any alteration would have spoiled its ability to survive. We know today that most organisms are not so perfectly adapted, and that most can tolerate considerable change before succumbing or, in rare cases, evolving into a new form adapted for a different way of life.

Working on the staff of the Museum of Natural History in Paris, Cuvier enlarged the museum's fossil collection from a few hundred skeletons to some 13,000. He saw that the deeper the stratum from which a fossil came, the more it differed from living animals, an observation that would later help paleontologists (students of fossils) relate changing skeletal forms to successive periods in the history of the Earth.

Cuvier also rebelled against the ancient doctrine of the "great chain of being," or *scala naturae*. This is the idea that all species can be arranged along one line, from the simplest microorganism to human beings. Cuvier insisted there were several major animal groups (vertebrates, for example, could not be on the same scale with mollusks, nor could either group be on the same scale with arthropods such as insects and crabs). He saw each group as its own Divine creation. Cuvier's groups have survived as today's phylums in the system of zoological classification, without, of course, the idea of separate creations for each group.

Cuvier also produced the first evidence that major groups of animals had become completely extinct. He explained this, as well as other changes from stratum to stratum in the crust of the Earth, as the result of no fewer than twenty-seven sudden, catastrophic events in the history of life on Earth, of which Noah's flood was only the latest. After each catastrophe, Cuvier held, God had repopulated the Earth with new species. With this view, Cuvier helped the Church to cling to a form of biblical literalism in the face of mounting fossil evidence that species had changed from one kind to another.

Cuvier's "catastrophism" has been replaced by "uniformitarianism," first articulated by Charles Lyell, the English geologist. This is the view that changes in the earth's surface are the result of currently active but slow geological processes that have been working uniformly for a very long time.

See EVOLUTION; LAMARCK; LYELL; UNIFORMITARIANISM; TAXONOMY.

DALTON, JOHN, 1766–1844, English chemist—John Dalton—the architect of the modern atom theory, which revolutionized chemistry and physics by setting forth the concepts of elements, atoms, molecules, and compounds—started life as a poor, physically awkward, personally unpopular boy who nonetheless had a bent for science and tried to share it with his community.

Young Dalton, the son of an impoverished hand-loom weaver in the tiny English country village of Eaglesfield, was recognized early on in his village as a mathematical wizard, and at the age of twelve he was allowed by local officials to start his own school. Unfortunately, Dalton was such an unappealing lecturer that he drew few students, and his little school had to close after three years because all the students had withdrawn.

At about the same time, Dalton began to show his bent for science by beginning to keep records of local weather conditions. He made his own instruments and recorded temperature, humidity, and wind and cloud conditions. It was a habit he would maintain without fail every day of his life. His notebooks contain more than 200,000 meteorological notations, the last of them entered on the day he died at the age of seventy-eight.

When young Dalton's school closed, he went to another village and joined his brother, who was already teaching school there. For twelve years Dalton taught in the school, continuing his own studies in Latin, Greek, mathematics, and science. Once again, Dalton tried to reach out to the community. He started something called the Science Discussion Forum, but had such a poor speaking voice and such an irritating manner on the platform that, again, attendance dwindled and the venture failed.

During this period, Dalton formed a crucial friendship with John Gough, a scholar who shared Dalton's interest in weather. Gough, blind from birth, spoke several languages and knew every species of plant in the region by touch, taste, and smell. Perhaps because of his blindness, Gough was not put off by Dalton's ungainly appearance and unappealing manner. Gough encouraged Dalton to publish his meteorological observations, and as a result Dalton was invited to join the Manchester Literary and Philosophical Society. Though his fame would eventually spread throughout Europe, Dalton maintained his association with the Manchester Society and insisted upon reading all of his more than a hundred scientific papers before this group even though it was far from England's brighter centers of scientific society at Cambridge or London.

The publication of Dalton's meteorological work also led to his being given a job at the age of twenty-seven as an instructor at a college in Manchester. He soon quit, however, because the administrative chores left him little time for science. He earned money by tutoring and, by living modestly and alone, he was able to carry out the research that would enshrine him as one of science's greatest practitioners.

Dalton published his sweeping new theories in 1808, and his ideas were adopted with surprising speed. The French elected Dalton to their Academy of Sciences and feted him upon his visit to Paris. The English Royal Society awarded him a medal. And in what was perhaps the most awkward matter for the country-born scientist, he was presented to the king, a ceremony that demanded Dalton wear knee breeches, buckled shoes, and a sword—all items forbidden to Quakers, which Dalton was. Dalton escaped by opting instead to wear the university robes he had received shortly before when accepting an honorary degree from Oxford University.

In 1844 Dalton died. The awkward country boy who couldn't get students to come to his school had, in fact, taught the entire world. Some 40,000 people filed past his coffin as it lay in state.

See ATOM THEORY; ELEMENT; ATOM; MOLECULE.

DARWIN, CHARLES, 1809–1882, English naturalist— Charles Darwin was the first "creation scientist." The father of evolutionary theory, the man once described as "the Newton of biology" and as "the most dangerous man in England," the man religious fundamentalists have vilified for more than a century, Darwin actually began his scientific career intending to demonstrate the glory of God's handiwork. "I did not in the least doubt the literal truth of the Bible," Darwin said of his beliefs as he set off, at the age of twenty-two, on his epic voyage aboard H.M.S. *Beagle.* The *Beagle* was a surveying ship, under the command of the deeply religious Captain FitzRoy, who had picked Darwin to be the ship's naturalist not because of his scientific acumen (Darwin was still an unknown), but because FitzRoy wanted a gentleman along for conversation.

For five years Darwin explored the wet jungles of South America and the dry rocks of the Galapagos Islands, looking, collecting specimens, taking notes, and trying to make sense of it all. The voyage gave rise to doubts about literal truth of Genesis and led eventually to the theory that propelled Darwin to scientific immortality.

Contrary to common belief, Darwin did not develop his theory of evolution during the voyage. He came to his ideas slowly, and to publishing them almost not at all. It was six years after Darwin returned to England (he would never again leave the country) before he wrote a brief outline of his ideas. It ran to thirty-five pages, but he was not ready to publish it. He wanted to amass more data. Two years later he expanded the sketch to 230 pages, but still he would not publish it, although by now he had published a journal of his voyage and had garnered a considerable following among English scientists.

Then, in 1856, Darwin learned that a young, unknown naturalist named Alfred Russel Wallace was independently coming to the same conclusions Darwin had harbored. Darwin's scientist friends warned him that if he didn't publish soon, his thunder would be stolen. He set to work on his magnum opus. Two years into the project and far from its completion, Darwin was shaken to receive in the mail a crisply written, 4,000-word exposition of the theory of natural selection, written by Wallace. "All my originality," Darwin said, "seemed smashed—and after twenty years of marshalling evidence."

Not wanting to see Darwin slighted, the great naturalist's friends arranged a joint reading before the prestigious Linnaean Society of one of Darwin's earlier, unpublished papers along with Wallace's paper. The claim that Darwin actually stole his ideas from Wallace is wrong. Darwin abandoned his "big book," as he called it, and reduced the project to a 490-page tome that he considered a mere abstract. *On the Origin of Species* was published in 1859, and all 1,250 copies sold out the first day.

Although the church attacked Darwin viciously, the theological case against evolution was largely withdrawn within Darwin's lifetime. He died knowing that the major Christian denominations had declared it possible to be both a Christian and a Darwinist.

Darwin was born to a moderately wealthy English family. As a child he was considered mentally inferior, and his family despaired of his ever amounting to anything. Until he joined the *Beagle* expedition, he spent most of his time drinking, hunting, and playing cards with classmates at Cambridge University, where he was studying for the ministry.

Upon his return to England, Darwin married his first cousin, Emma Wedgwood, granddaughter of the pottery maker. He lived in virtual seclusion at Down House in the countryside of Kent, seeing no one but his family and close scientific friends. For forty years he kept to a clockwork

schedule, interspersing walks around his private grounds
with scientific work in his study and—his great pleasure—backgammon games by the thousand with Emma.
He wrote many books, all of them commercial successes
by the standards of the times. They ranged from his first,
the chronicle of the voyage, to the last, *Formation of Vegetable Mould Through the Action of Worms*. Presaging
the field of sociobiology, which in the 1970s would assert
that behavior, both animal and human, was shaped by
evolution, Darwin even wrote a book on how human emo-
tions are governed by evolved traits of the brain.

Darwin died of a heart attack at the age of seventy-
three and, though he was the arch unbeliever of his age,
England laid him to rest in the consecrated ground of
Westminster Abbey.

See EVOLUTION; LYELL; NATURAL SELECTION.

DAVY, HUMPHRY, 1778–1829, English chemist—
Though his name often appears in lists of scientific greats,
Davy was not really responsible for any deep new insights
into nature. He discovered the elements sodium and potassium, and was an ardent exponent of the scientific
method in a time when it was only first being established.

Davy's prominence in the lists of renowned persons may
be owing to the fact that he mixed almost as much with the
artistic and literary set in England as with his scientific
colleagues. His contacts among the famous began when he
discovered the intoxicating effects of inhaling nitrous oxide, the so-called laughing gas that dentists use today. It
quickly became fashionable to visit Davy's laboratory to
"take the airs." Davy was also something of a poet and,
through his offers of drugs (as some today might label nitrous oxide), he met the poets William Wordsworth, Sam-
uel Taylor Coleridge, and Robert Southey. In his later life,
Davy was knighted and was married to a wealthy widow.

Davy's scientific reputation is somewhat marred by his
stout resistance to John Dalton's atomic theory, the beginning of our modern concept of atoms.

See ATOM.

DECAY (ORGANIC)—This is a fairly vague term in biology, referring to almost any kind of breakdown of once-living tissues and their constituent molecules. Unlike radioactive decay, organic decay, or rotting, is not a spontaneous process. It is the work of various small organisms, ranging from certain insects and worms to microbes such as bacteria and fungi, that literally eat the dead tissue and digest it into smaller components. Thus decay and digestion—even inside the human intestine—are much the same process. Decay is absolutely essential to the maintenance of life on Earth, for without it, all the available molecules needed by organisms would quickly become locked up in an accumulating global litter of dead plants and animals.

See PASTEUR; REDI; SPALLANZANI.

DECAY (RADIOACTIVE)—The nuclei of some atoms, such as uranium and radium, are naturally unstable, and from time to time will spew out one or more of the particles of which they are made (neutrons and protons) or emit a burst of electromagnetic radiation. Left to its own devices, such an atom has a 50–50 chance of doing this within a period of time known as its half-life. The half-life of the most common form of uranium is 4.5 billion years, and of the most common form of radium, a mere 1,622 years. Odds are that half the atoms of a given element will decay within the half-life. Some forty naturally occurring substances are radioactive. Many other elements can be made radioactive by bombarding them with subatomic particles that enter the nuclei of the atoms of these elements, making these nuclei unstable and causing them to decay with the emission of energized subatomic particles. This is why the materials surrounding the fuel in a nuclear power plant become radioactive.

In a nuclear power plant, radioactive decay is speeded up beyond its natural rate when neutrons emitted by one decaying atom are allowed to penetrate another, potentially radioactive atom. Without this, the decay that might not occur for millions of years can be made to happen immediately.

See ATOM; ATOMIC STRUCTURE.

DEMOCRITUS, 460 B.C.–370 B.C., Greek physicist, mathematician—Although Democritus is often named as the inventor of the theory of atomism—the idea that everything is made of invisibly small particles— he actually got the idea from Leucippus, one of his teachers. Democritus merely became the theory's greatest proponent. *Atomos* is the Greek word meaning "indivisible." Democritus thought that if you cut up objects finely enough, you would theoretically come down to the atoms—hard, indivisible particles—of which they were made.

Democritus said that there might be an infinite number of kinds of atoms, differing in shape and size. People, he said, were made of a mixture of body atoms and soul atoms. Soul atoms, like fire atoms, were spherical. Moreover, soul atoms were what give the body atoms life. Democritus taught that although a person's soul did not survive death, its atoms could not be destroyed. Eventually, they were recycled and found their way into new living beings.

Though it is tempting to poke fun at it in hindsight, Greek atomism was a stunning insight, for there is, in fact, no easy evidence to support it. It was simpler then to think of matter as a kind of continuous quantity than as something made of little, indivisible lumps. Although atomism was taught in Greece until the time of Aristotle, it was thereafter largely ignored until the time of Galileo. About then it came to be accepted in a general way, partly because of the observations of alchemists that chemicals combined in fixed ratios—a phenomenon which implied that matter existed in particles that combined only in units of whole numbers (not fractions). Still, it was not until the nineteenth century that experimental evidence of atoms began to appear. The final proof was nailed down by none other than Albert Einstein in one of the three great papers he published in 1905. (Relativity was only one of the three.)

Although the word atom is still used, its meaning is anything but the one intended by Democritus. Atoms, we now know, are divisible. Moreover, the smaller particles that make atoms (electrons, protons, and neutrons) are themselves divisible into still smaller particles called quarks.

See ATOM THEORY; EINSTEIN.

DESCARTES, RENÉ, 1596–1650, French mathematical philosopher, scientist—Descartes was the first modern philosopher of science. It was he who proposed that all of nature, including the bodies of human beings, could be explained through purely mechanical processes—the actions and reactions of objects and energies mindlessly obeying rational physical laws of cause and effect. The French philosopher made only one exception to this—the conscious human mind, which he felt had some mystical qualities. Descartes's view was a radical change from the long-held position that magical and supernatural phenomena lay behind practically everything that happened in nature. By declaring the world to be essentially a mechanistic place, Descartes opened all of nature to scientific scrutiny.

Descartes was the son of a lawyer and a wealthy mother who died shortly after his birth, leaving him enough money to make him financially independent for the rest of his life. He was educated in Jesuit schools, earned a law degree, and spent several years traveling throughout Europe and in voluntary military service, although there is no record of his doing any real fighting. While in the military, he got the idea of inventing his own philosophical-intellectual system. Seeking more peaceful circumstances in which to develop his system, he quit the army at the age of thirty-two and moved to Holland, where the intellectual environment was freer.

Descartes did a considerable amount of experimentation, dissecting animals and their embryos, measuring the weight of air, observing the motions of vibrating strings and the optical effects that explain rainbows. Perhaps his most useful contribution was the invention of what is known today as the Cartesian coordinate system. This is a type of geometry in which the properties of various shapes and curves can be described by algebraic equations, and studied by subjecting the equations to purely mathematical treatment.

In 1649 Descartes was enticed by the Queen of Sweden to tutor her in philosophy. He went to Stockholm and, according to the queen's dictates, met with her every morning at five, no matter how icy the Swedish winter or how poor the heat in the castle. Descartes, who had made a lifetime habit of spending his mornings doing his best thinking under the covers of a warm bed, caught pneumonia. The queen's doctors zealously bled him and, between the infection and the bleeding, Descartes died. He was just fifty-three.

DNA—This is a chemical that can store information. Living organisms rely on the information stored in DNA to control how they grow from a single cell to a complex, fully developed adult. The information in DNA tells each cell what specialized features to develop (making one cell a nerve cell, another a liver cell, and so on), and what ongoing processes to engage in. DNA, in other words, is the master molecule of life.

DNA is an abbreviation for DeoxyriboNucleic Acid, a molecule consisting of a pair of long chains of smaller molecules, the two chains being bridged by molecules acting like the rungs of a ladder. The vertical parts of the molecule are always the same; the rungs are the interesting parts because there are four different kinds, and the sequence in which they are stacked is the way messages are encoded in DNA. As it happens, the DNA ladder is twisted into a kind of spiral staircase. Each of the two vertical parts of the ladder has the shape known as a helix, and the two together are therefore known as a double helix. DNA normally stays in the nucleus. Its message is transported to the rest of the cell by a similar kind of molecule called "messenger RNA."

See next entry, DNA, HOW ITS ROLE WAS DISCOVERED; DOUBLE HELIX.

DNA, HOW ITS ROLE WAS DISCOVERED—As far back as 1866, Gregor Mendel's pea-breeding experiments in his Austrian monastery garden proved that inherited traits are governed in living organisms by discrete units, and not by a blending of bloods from two parents, as many people believed. But Mendel had no idea what the units were. They remained strictly a concept represented by no known physical object. Other scientists named the units genes in 1909, but still they remained ignorant of what a gene actually was. It was not until the 1940s that biologists figured out that genes were made of DNA, a chemical that had been known since the nineteenth century, and that these pieces of DNA resided in some curious structures within the cell nucleus called chromosomes. The word chromosome means "colored body," and was merely a reference to the fact that laboratory stains could make these structures look colorful. Not until biologists knew exactly which molecules were involved in transmitting hereditary traits could they begin the basic research that has led to today's explosion in genetic engineering and the promise of curing genetic diseases.

Mendel's findings drew little attention in his own time, and fell into obscurity until the beginning of this century. One reason for this may have been the statistical nature of his evidence, an approach that would not be appreciated until fairly recently. In 1902 a Columbia University student named Walter S. Sutton had the good fortune to read Mendel's paper during the period in which he was observing grasshopper chromosomes, and he noticed that they fit many of Mendel's observations about units of heredity. For example, both chromosomes and Mendel's hereditary units come in pairs. The pairs split up when an organism makes its reproductive cells (sperm or eggs), with one member of each chromosome pair going into each cell. When a sperm fertilizes an egg, the resulting, new individual again has paired chromosomes.

Since no other structures in cells could be seen to behave this way, Sutton guessed that Mendel's units of heredity must be in the chromosomes. The proof would be some years in coming, but Sutton was right, and Columbia became a center of early genetics research.

Sutton did no experiments to prove his guess. That step fell to another Columbia researcher, Thomas Hunt Morgan, who selected fruit flies as the animals in which to study chromosomes. It was a choice that has benefited generations of geneticists since. The fruit fly is still a favorite subject because it is small and easy to keep in the laboratory, produces several generations a month (making breeding experiments short), has only eight chromosomes (humans have forty-six), and has salivary glands whose cells for some reason contain giant chromosomes some 200 times larger than those in the fly's other cells. (The genes are the same in all of the fly's cells; it is just that the giant chromosomes have more non-genetic structural material in them.) Through thousands of breeding experiments, Morgan was able to sort out the inheritance patterns of scores of the fruit fly's physical traits (eye color, wing shape, and so forth), and to establish that the genes must indeed be in the chromosomes. What he found, basically, was that certain traits were likely to be inherited in groups, and that certain groups were more likely than others to reappear in progeny. Morgan's statistical analysis led him to conclude that this could only be so if the genes determining those traits were somehow tied together on the chromosomes, and not only tied together, but tied together in a linear fashion like links on a chain or beads on a string. Morgan won the Nobel prize in 1933 for this work.

Still, however, nobody knew what genes or chromosomes were made of, and until their chemical nature could be understood, how they might work was a total mystery. There were two leading ideas about this—that genes were made of special protein molecules, and that they were made of deoxyribonucleic acid, or DNA. DNA was a candidate because it was found only in the cell nucleus, where the chromosomes were. But as of the 1940s, most scientists discounted DNA because it seemed to be too simple a molecule to do all the things genes must do. DNA was nothing more than a long chain of subunits, and there were only four kinds of subunits in it. Geneticists knew that genes, whatever they were, had to somehow encode lots of information, and a code with only four symbols, such as that in DNA, didn't seem too versatile. Proteins, on the other hand, were enormously complex structures made out of twenty kinds of subunits that could be put together with almost endless variations of shape and ability to react chemically. The twenty subunits of proteins are almost as many as the twenty-six subunits of the English alphabet, which can be used to encode an uncountable number of messages. A mere four-letter alphabet didn't seem terribly useful.

The decision, of course, eventually went to DNA, but not until after acceptance of a long and involved body of work dealing with quite another problem. The first part of the work was done by Frederick Griffith, a British researcher working on pneumonia in the 1920s. He was studying a pneumonia bacterium called pneumococcus, and knew that it came in two forms—one that had a smooth coat under the microscope and one that had a rough coat. The smooth-coated bacteria caused disease while the rough-coated ones didn't. Griffith's practice was to inject the bacteria into mice to determine whether they caused disease. When mice were given the harmless bacteria, they stayed healthy. The harmful bacteria gave them pneumonia. If Griffith killed the harmful bacteria first and then injected them, the mice stayed healthy. Dead bacteria can't cause disease. One day, however, Griffith tried something different. He injected mice with live harmless bacteria and dead harmful bacteria. To his great surprise, the mice developed pneumonia, and when he examined their lungs, he found live, smooth-coated harmful bacteria. Somehow the dead harmful bacteria had transformed their harmless cousins into killers. Moreover, the harmless bacteria had acquired the ability to produce the smooth coat of their harmful cousins.

Griffith died never knowing the full significance of his

experiment. Researchers in several countries repeated
the experiment and got the same results. Some also found
that injecting the bacteria into mice was unnecessary. Put
dead smooth bacteria in a test tube with live rough bacteria, and before long the live ones will acquire the smooth
coat. In 1932 came one more development. Several scientists joined the hunt for the transforming factor—whatever it was—that the live bacteria got from the dead
bacteria to acquire their traits. In 1932 one scientist took
the dead bacteria, mashed them up, and forced them
through a very fine filter. Large fragments were strained
out and only a clear fluid came through. The fluid, it
turned out, was just as good at transforming live bacteria
as were whole dead bacteria. The transforming agent appeared to be some substance in the fluid.

At this point Oswald Avery, a doctor doing research at
New York's Rockefeller Institute, picked up the trail and
followed it for ten years. He took the clear fluid, which
was actually a broth consisting of thousands of kinds of
proteins, fats, starches, sugars, and DNA. He laboriously
separated out the various constituents, purified them, and
tested them to see whether they would transform live bacteria. Eventually Avery got down to DNA, and sure
enough, that alone could cause the transformation. To
check this discovery, Avery prepared more DNA and
added an enzyme that does nothing except break down
DNA. This concoction, lacking undamaged DNA, failed to
transform bacteria. DNA, Oswald Avery concluded in
1944, must be the substance that carries the gene or genes
for a pneumococcus's smooth coat and its disease-causing
ability.

Other researchers following other approaches would
come to the same conclusion over the next few years,
amply confirming Avery's discovery. The transfer of one
bacterium's genes to another seemed a strange thing back
then, but today it is known that bacteria regularly come
together, and in a sense mate, transferring genes from
one individual to another. The process is called recombination, and it is what led to today's so-called Recombinant

DNA technology in which biologists transfer genes among
organisms. In the years after Avery's discovery, it became clear that DNA, a seemingly simple kind of molecule, encoded the genetic message. But how? Cracking the
code would be the next big achievement.

See MENDEL; GENES, HOW THEY WORK; DOUBLE HE-
LIX; RECOMBINANT DNA.

DOBZHANSKY, THEODOSIUS, 1900–1975, natu-
ralized American biologist. See MODERN SYNTHESIS.

DOPPLER, CHRISTIAN J., 1803–1853, Austrian
physicist. See BIG BANG.

DOPPLER EFFECT—This is a phenomenon in which
the perceived pitch of a sound or the frequency of an electromagnetic radiation (such as light) changes if the sender
and the receiver are moving with respect to one another.
The best known example involves the sudden change in
pitch of a train whistle as the locomotive passes a stationary listener. When the train is approaching, the whistle
has a higher pitch than when it is receding. A passenger
on the train may experience the same effect when listen-
ing to a bell at a highway crossing. The tone suddenly
drops to a lower pitch when the passenger passes the bell
and is moving away from it. The Doppler effect, named
for Christian Doppler who described it in 1842, is the
cause of the so-called red shift in the light emitted from
distant galaxies. Measurements of the degree of this red
shift (a reduction in the perceived frequency of light
waves or movement of their frequency toward the red, or
low-frequency, end of the spectrum) have led to the conclusion that the universe is expanding. (See BIG BANG.)

The Doppler effect works on any phenomenon that occurs in the form of waves, but is probably easiest to visualize by imagining a gun that fires one bullet every second
at a target. When gun and target are stationary, the bullets hit the target once every second. Now imagine the
gun mounted in a truck that is speeding away from the

target. Every time a bullet is fired, the truck is a little farther from the target, which requires the bullet to travel farther and takes it longer before it hits the target. As a result, an observer at the target would notice that the time between bullets is longer than a second. The bullet frequency would be shifted to the slow end of the range of frequencies (analogous to the low-frequency, or red end of the light spectrum that astronomers refer to when they speak of a red shift). On the other hand, if the truck is speeding toward the target, the distance diminishes with each bullet, and the bullets hit the target more frequently than once a second. In either case, there has been no change in the true rate of firing, only in the perception of that rate.

In the same way, the Doppler effect distorts the perception of any kind of energy that is emitted in a wave or pulsed form. Take sound as an example. Middle C on the piano is a tone with a frequency of 440 hertz (cycles per second). In other words, the piano string vibrates 440 times per second. The vibrating string makes the surrounding air vibrate at the same rate. If both piano and listener are stationary, the vibrations traveling out from the piano at the speed of sound will hit the eardrum 440 times a second. But if the listener runs toward the piano, the distance from sender to receiver will be diminishing, and the listener's ear will intercept more than 440 vibrations in a second. The note will sound too high-pitched. Conversely, if the listener runs away, it will take the sound waves slightly longer to reach his ear because he is steadily increasing the distance they must cover; hence he will hear fewer vibrations in a second, producing a lower tone.

DOUBLE HELIX—The discovery of the double helix or spiral staircase structure of DNA in 1953 has been called the greatest advance in biology since Darwin. That may be an overstatement, but it is unquestionably one of this century's most brilliant achievements, for it explains one of the major wonders of life—how DNA replicates itself

HOW DNA REPLICATES

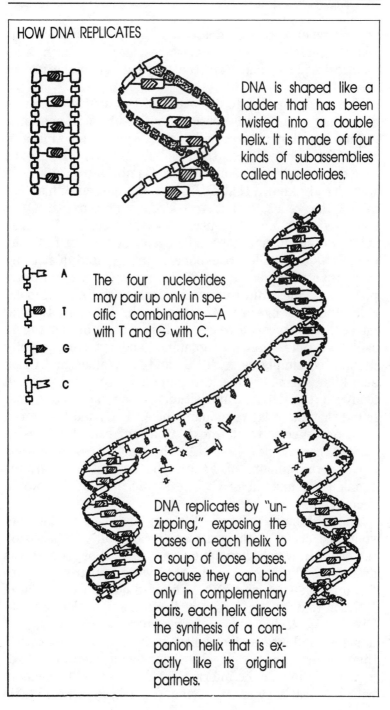

DNA is shaped like a ladder that has been twisted into a double helix. It is made of four kinds of subassemblies called nucleotides.

A

T

G

C

The four nucleotides may pair up only in specific combinations—A with T and G with C.

DNA replicates by "unzipping," exposing the bases on each helix to a soup of loose bases. Because they can bind only in complementary pairs, each helix directs the synthesis of a companion helix that is exactly like its original partners.

so that when a cell divides, each of two "daughter" cells has a complete set of chromosomes, each of which is an identical copy of one from the original set in the parent cell.

The work that led to discovery of the double helix was done chiefly by Francis Crick, a British physicist specializing in crystal structure, and James Watson, an American geneticist. They began their collaboration at England's famed Cavendish Laboratory in Cambridge in 1951. It was already known that DNA was the molecule that contained the genes, and it was already known that DNA consisted of a monotonously repetitive chain of similar subunits, small molecules called nucleotides. In fact, the chain was a so-called backbone that consisted of nothing more than a long sequence of alternating sugar and phosphate molecules. But from each sugar a third kind of molecule stuck out like a tooth on a comb. These protruding molecules were called bases, and they came in four types, called adenine, thymine, guanine, and cytosine, usually designated simply as A, T, G, and C. Watson and Crick also knew that an American named Erwin Chargaff had analyzed the DNA composition of several species and found that the relative amounts of A, T, G, and C differed from one species to the next. Chargaff had also found a curious consistency in DNA: the amount of A always equaled the amount of T, and the amount of G always equaled the amount of C. The significance of this would become apparent later.

The full story of how Watson and Crick (the conventional order of their names, incidentally, is based on the order in which they appear on the 1953 paper and was determined by a coin toss as they were preparing to publish on the structure of DNA) figured out the structure of DNA is told most candidly in Watson's now classic book, *The Double Helix*. It is a story of personal rivalries, wild hunches, and intense jealousies on the part of people other than Watson and Crick whose hard-won laboratory findings gave the pair the information they needed to come up with the final answer. Watson and Crick, in fact, did little laboratory work themselves. For the most part, they simply got together and talked about how the various findings concerning DNA, like Chargaff's nucleotide ratios, might be represented in a DNA structure. Watson and Crick spent most of their time building models—physical objects that looked a bit like Tinker Toy constructions, with balls representing atoms and sticks for the bonds between them. Some of their models were simple cardboard cutouts of adenine, thymine, guanine, and cytosine molecules that they tried to fit together in various ways.

One of the most important pieces of laboratory data about DNA came from Rosalind Franklin of King's College, in London. She had made x-ray pictures of DNA crystals, but since such pictures are not as easy to interpret as x-rays of bones, it took considerable ingenuity to figure them out. Her x-ray crystallographs, as they are called, showed that DNA was not one chain but two chains running in opposite directions. Additional information came from chemists who told Watson and Crick that adenine would naturally form a weak bond with thymine, and cytosine with guanine.

After a year and a half of playing around with models, trying to make them fit all the data that was known about DNA, Watson and Crick finally hit on a model that agreed with the data. In this model, the nucleotides in each of the two chains of DNA are securely linked to one another by the sugar-phosphate bonds of the backbone. And the four kinds of bases that stick out are securely bound to the backbone. The two chains are bound to each other by weak bonds between the bases. If, for example, the sequence A-C-T-G is found on one chain (along the backbone the bases may be in any order), the opposite chain would have to have the sequence T-G-A-C. This is because T binds only to A and G only to C.

Once Watson and Crick built their model according to these rules, they found that the only way to make the bases come close enough to form the weak bonds that held them to one another was to make each backbone into a spiral. This rotated the teeth of the comb—the bases sticking out from the backbone—neatly into the proper position to bond to one another. They called the paired spirals a double helix.

The model immediately told the scientists how DNA can duplicate itself. Since the bases are only weakly bound, they can become unbound as the flexible but secure backbones unwind from one another. Free nucleotides floating nearby could then automatically come into position along each single backbone, binding only where a complementary base occurred on the backbone. In other words, a T that had broken away from an A as the backbones separated would soon be replaced by a new T. As the double helix "unzipped," each chain would serve as the template for the assembly of its complementary, opposite number. If the entire DNA helix became unzipped, each chain would direct the construction of its missing partner. Watson and Crick's design for DNA was a stunning demonstration of the power of molecular modeling. The elegance of the model immediately captured the scientific community, and in laboratories all over the world researchers confirmed it in a variety of actual laboratory experiments beyond the scope of this book. For their work, Watson and Crick won the Nobel prize.

See WATSON; CRICK; DNA; GENES, HOW THEY WORK.

EARTH, ORIGIN OF—Our planet did not have an independent origin. It was one of several bodies that formed as part of a much larger process that created the whole solar system from a single giant cloud of gas and dust.

As the sun was forming at the center of this swirling cloud, colder particles of gas and dust clumped together in trillions of places around the sun. At first, electrostatic charges made the particles stick together, but as they grew, the influence of gravity came to play a more powerful role. Over many millions of years, the sun continued developing, and countless tiny planets, or planetismals, formed. Gradually the larger planetismals, through their gravity, pulled in the smaller ones. During a period of perhaps 100 million years, the clumps of matter grew.

The clump that was to become the Earth was at first cold, and on its surface, particles simply piled up on one another. As the clump grew in mass, however, the pressure at the center increased, and with it, so did the heat. As meteorites collided with the growing planet, their energy of impact was also converted to heat. When the heat became enough, it melted the particles that had piled up, and most of the heavier atoms, such as iron, sank to the core of the planet. Lighter atoms, such as those of gases, floated outward to form a blanket over the solid planet. Lighter solids, such as the silicates that make rock, floated atop the molten core. By about 4.5 billion years ago the Earth had been born as a fully differentiated planet.

See UNIVERSE, ORIGIN OF; SOLAR SYSTEM, ORIGIN OF.

EARTHQUAKE—The tremors that rumble through the Earth's surface are only the merest twitchings of a planet deep in subterranean turmoil. Yet these twitchings are the most obvious demonstrations we have that the continents are moving, riding the huge plates that make up the Earth's fractured crust as they bump and grind against one another.

Dig down deep into the Earth's crust and the temperature gradually gets warmer. The deeper you go, the hotter it gets. The hotter rock gets, the softer it becomes, until, when it is red hot, it is a thick, viscous fluid and will slowly flow. Rock can even get white hot and become as thin as oil. The Earth's crust is solid only for about 20 miles down. Then it gradually begins to soften until, 200 or 300 miles down, it is fairly viscous. Down in the molten regions of the Earth there are currents flowing ponderously. Hotter rock from deeper down rises, reaches the harder, cooler rock above, and is deflected to one side. As this sideways-flowing molten rock cools, it sinks, maintaining the cycle.

Floating atop all this heavy, slow-moving material are the rigid plates of the Earth's crust. The movement of the soft rock below literally pushes these crustal plates

around. Where one plate touches another, the two may
grind or collide head on, depending on the patterns of currents below. Since hard rock cannot yield gradually to the
forces pushing the plates, they build up over a long period
until the contact between the plates breaks and one
lurches against the other. The break and the lurch are felt
as an earthquake.

Almost all the world's earthquakes occur along the lines
joining one crustal plate with another. California's San
Andreas fault is, in fact, where the North American plate
meets the Pacific Ocean plate. People living west of the
fault are, geologically speaking, not living in North America but on the upraised edge of the Pacific plate. The Pacific plate is being pushed northwestward, relative to
North America, and every quake along the San Andreas
fault sees a piece of the West Coast lurch toward Alaska.

See PLATE TECTONICS; WEGENER (the originator of
plate tectonic theory).

ECOLOGY—"The ecology" is not something out there in
the forest or in an estuary. Ecology is the science that
studies forests, estuaries, and all other habitats, relating
the various components of the environment to individual
species of plants and animals. Ecology seeks to learn the
relationships among the various living and nonliving elements in an ecosystem, or habitat. The term "ecology" was
coined more than a century ago by a German biologist, long
before the environmentalist movement developed.

ECOSYSTEM—In the deepest sense of the term, there
is only one ecosystem—the Earth as a whole. However,
like all scientists, ecologists find it more convenient to
subdivide their territory into smaller units that are more
or less homogeneous. Thus ecologists may speak of a de-
ciduous forest ecosystem or a coral reef ecosystem. The
term is more or less synonymous with habitat.

EHRLICH, PAUL, 1854–1915, German medical researcher—Most of the patients whose infectious diseases

are treated today with drugs owe a debt to Paul Ehrlich. It was his idea, around the turn of the century, that certain chemicals taken into the body might kill or cripple only the invading microbes while leaving normal cells alone. Ehrlich referred to such chemicals as "magic bullets" that invariably aimed themselves at the target. Though the concept still tantalizes drug researchers, it has turned out that few if any drugs act only on their target cells, virtually all also act on other cells, producing side effects.

Ehrlich, who was born into a prominent industrial family in Prussian Silesia, which is now in Poland, sought his first magic bullet in the treatment of syphilis. He started with the observation that certain chemical dyes stained certain parasitic organisms more readily than they stained ordinary human cells. The dyes were used to make the organisms more easily visible under a microscope. The dyes had their staining effect because they somehow bound themselves to the molecules of the target cells. If so, Ehrlich reasoned, they might also be made to disrupt the functioning of those cells, perhaps preventing them from multiplying while leaving normal cells unaffected.

Ehrlich focused on the microscopic organism, called a spirochaete, that causes syphilis. He started with the dyes microscopists used to view this organism, and made hundreds of slight chemical modifications to the dye molecules. Each new chemical was tested as a stain on spirochaetes and human cells, always looking for a substance that strongly stained the spirochaetes but did not stain human cells at all. Ehrlich made and tested some 3,000 compounds, and the best turned out to be number 606, a compound of arsenic. Ehrlich tried it in human syphilis patients and it worked. After many further tests to assure himself that the arsenic compound was safe, it was released as the first truly useful treatment for syphilis. Worldwide honors were heaped on Ehrlich. In modern times Ehrlich's chemical, marketed under the trade name Neosalvarsan, has been superseded by antibiotics.

EINSTEIN, ALBERT, 1879–1955, German-Swiss-American physicist—Albert Einstein, arguably the greatest scientist since Newton 300 years earlier, had probably the least auspicious start in life. At first he showed signs of mental retardation, not beginning to talk until he was three years old and not talking fluently until the age of nine. He hated school, got poor grades, and as a teenager was expelled for being a "disruptive influence." Einstein flunked his first college entrance examination, passed on the second try, got mediocre grades and, after graduation, could not find regular work. Largely unemployed for the next two years, he survived by working at odd jobs as a tutor until he finally landed a full-time position with the government, checking patent applications for technical accuracy.

In his spare time Einstein worked on his hobby—devising a revolutionary new way of understanding the nature of matter and energy, of time and space, or, to be concise, a new way of making sense of all existence. In 1905, while still working at the patent office, he wrote up some of his new ideas as four articles, and sent them off to a physics journal. Though Einstein had no credentials and was only an amateur, the articles impressed the editors and were published in a single issue.

The first paper suggested a way of proving once and for all whether matter was really made of tiny particles, or atoms, or was not. Experiments later confirmed Einstein's idea that it was, and yielded the first direct evidence for the reality of atoms. The second paper, building on the ideas of Max Planck, German physicist, suggested that electromagnetic radiation, a fundamental form of energy pervading the universe, actually consisted of discrete particles called quanta or, later, photons. Einstein, again shown correct, eventually received a Nobel Prize for inventing quantum physics in this paper. The third article, today the most famous, set forth the special theory of relativity, which asserts that space and time are not absolute but vary with local circumstances, including the popular idea that time slows down for a moving body. The fourth paper introduced Einstein's idea that matter and energy are interchangeable forms of the same thing. It contained the simple formula $E = mc^2$, which means that the amount of energy, E, that you can get from matter is equal to the mass, m, of the matter multiplied by the speed of light, c, multiplied by itself, or squared (a very large number).

At a stroke, or at four simultaneous strokes, this twenty-six-year-old spare-time mathematician and hobbyist in theoretical physics altered forever the way in which human beings would envision the nature of the universe, from the smallest packets of energy inside atoms to the largest galaxies and most distant quasars. By revealing the ultimate equivalence of matter and energy, Einstein also opened the door to the atomic bomb and the hydrogen bomb.

"The world," said Aldous Huxley after hearing of Einsteinian physics, "is not only queerer than we imagine, it is queerer than we *can* imagine."

Albert Einstein was born in the small German town of Ulm. The year after he was born the family moved to a suburb of Munich. There Einstein's father owned and operated a small electrochemical plant. Einstein's bachelor uncle, an engineer, worked in the plant and lived with the family, and apparently introduced Albert to math. Einstein's mother was interested in music, and forced Albert to take violin lessons. Later he would recall that he hated the lessons, but he became a reasonably good violinist and continued to play as a form of relaxation for the rest of his life.

The Einsteins were Jewish but not religious. At that time all schools in Munich were run by religious groups, and the Einsteins simply sent Albert to the nearest one, which happened to be Catholic and to include religious instruction. In secondary school he received instruction in the Jewish religion. The contrasting religious educations left young Albert, he would later say, with a respect for the ethical values of religion but a conviction that religious rituals were based on superstitions that blocked independent thinking. After secondary school, Einstein gave up formal religious affiliations until the rise of the Nazi regime in Germany persuaded him to reassert his kinship with the persecuted German Jews.

While Albert was in secondary school the family business failed, and the Einsteins moved to Milan to make a new start. It was arranged for Albert to stay behind in school in Munich. At the age of twelve young Albert declared that he would devote his life to solving the riddle of the "huge world." At fifteen, however, he was expelled and went to rejoin his family. Young Einstein, thoroughly hating the lock-step German school system and other things German, persuaded his father to apply for a revocation of Albert's German citizenship. This was granted in 1896, and until 1901, when Einstein obtained Swiss citizenship, he was stateless.

Einstein had gone to Switzerland in the hope of finding a more congenial school system. He succeeded, completing secondary school there and taking his college education at the Swiss Federal Institute of Technology in Zurich. In 1903, a year after starting work at the patent office in Berne, Einstein married his college sweetheart, Mileva Maric. Two years later he published the four articles that would immediately transform not only physics but Einstein's life. Public recognition of Einstein's achievements would not come for some years, but the scientific community quickly took Einstein to its bosom. He got a series of professorships and then, to allow him more time for research, a position at the Prussian Academy of Sciences in Berlin that required only an occasional lecture at a nearby university.

Einstein spent most of his time working on a broader form of the relativity theory. The first version of the theory applied only to the special case of objects at rest or moving at an unchanging rate. To deal more completely with the real world, where objects are constantly changing speed with respect to one another, Einstein wanted to work out a so-called general theory of relativity. This he published in 1916, having worked it out in the midst of a Europe engaged in its first world war.

The war had a major influence on Einstein. It stranded his wife and two sons in Switzerland for several years, an enforced separation that was to lead to divorce. The war also delayed the first experimental test of his new theory (which depended on making an observation of a star near the sun during a solar eclipse). And the war rekindled Einstein's pacifism. He corresponded with leading political pacifists and developed a world view that would increasingly come to dominate his life. To some, Einstein's view of human nature seemed naive. Upon the armistice of 1918, for example, Einstein was totally convinced that militarism had been abolished in Germany.

In 1919, England's Royal Society announced that its expedition to observe a solar eclipse in the South Pacific had confirmed the theory of relativity. The theory predicted that starlight reaching the Earth on a path that grazed the sun would not follow a straight line but a curved line. Because the sun is so bright, the only time we can see stars that appear near it is during a total eclipse. The observations showed that a star which was in a certain position appeared to be shifted to one side by an amount that Einstein's equations had predicted. The path of its light had been bent. The finding captured the world's imagination, and Einstein was catapulted into international renown as the world's greatest scientific genius. He became a celebrity, hounded by photographers who delighted in his shaggy hair and rumpled suits, and by reporters, for whom it was all the better that his theory could only be understood by a few of the world's greatest minds. The fact that he was a theoretical physicist mattered not. Reporters eagerly sought Einstein's opinion on all sorts of matters, as trivial as baseball and as profound as the deteriorating political situation in Germany. On this latter question and its larger implications, Einstein was ready with his eloquent protestations on the horror of war and the need to avoid it.

For three years after the eclipse observation Einstein and his new wife, Elsa, traveled the world, lecturing on relativity, meeting with prominent intellectuals and pacifists, and sightseeing. So many people turned out for his lectures that an impresario once guaranteed Einstein a three-week booking at London's Palladium. He didn't sign but he did accept a request to tour the United States to raise money for Zionist causes.

Through all the tours and lectures and flashbulbs, Einstein kept trying to work on what he hoped would be his next big advance, a so-called unified field theory that would show gravity and electromagnetism to be different aspects of the same basic phenomenon. He published papers on this, but they were respectfully dismissed by his colleagues as entirely too preliminary or as frankly out of step with the growing implications of another major advance of twentieth-century physics, quantum theory. Although Einstein had contributed to its earliest stages with his paper on light particles, he rejected the statistical nature of quantum theory as others were developing it. For example, quantum theory insists that the behavior of individual subatomic particles cannot be predicted with certainty, but that only the behavior of groups of such particles can be predicted, and then only with a given probability of doing something. The very notion of this ran counter to Einstein's innate sense of how the world works. "God does not play at dice," he declared, dismissing the theory that almost every other physicist already knew to be as important as relativity. Einstein's refusal to accept quantum theory drove a wedge between him and his colleagues. "Many of us regard this as a tragedy," said Max Born, a German physicist and a close friend of Einstein's at the time, "both for him as he gropes his way in loneliness, and for us, who miss our leader and standard-bearer."

Einstein turned fifty in 1929, and his world began to crumble. A major paper on unified field theory was coolly received. Arabs attacked Jewish settlers in Palestine. The Nazis gained ground in Germany. The Great Depression began. And closer to home, Einstein's younger son Edward suffered a mental breakdown and blamed his father whom he said he worshiped from afar but who had abandoned his fatherly duties. Einstein's relations with his older son, Hans Albert, remained cordial. As the political situation in Europe deteriorated, Einstein devoted more and more time to espousing pacifism, joining various international disarmament committees and lecturing to audiences of all sorts.

In 1933, shortly after Hitler came to power, Einstein renounced his reacquired German citizenship and prepared to leave his homeland for good. His friends, fearing for his life, slipped him out of the country as quietly as possible, first to Belgium and then to England aboard a private yacht. Eventually Einstein reached the United States and accepted a position at Princeton University's new Institute for Advanced Study. Here he would remain for the rest of his life. He and his wife settled into a modest house and adopted a quiet routine that seldom varied. Einstein would walk the mile or so to his office at the Institute every morning, work on his theories there, receive visitors, and answer letters. His chief diversions were his violin and his sailboat.

In the fall of 1939 the winds of war were stirring in Europe. Einstein and his colleagues became aware that German physicists were on the track of experiments that could lead to an atomic bomb. Fearing that Hitler might discover a weapon of unprecedented power, a weapon that might fulfill his goal of world conquest, a group of physicists decided that the only course of action would be to beat Hitler to the bomb. They hoped to persuade President Roosevelt to launch the effort and came to Einstein, asking him to lend his prestige in attempting to persuade Roosevelt of the need for it. Though a pacifist, Einstein recognized the horror that could come if Hitler had his way, and wrote a letter to Roosevelt. "Some recent work by E. Fermi and L. Szilard which has been communicated to me in manuscript leads me to believe that the element uranium may be turned into a new and important source of energy in the immediate future," he said. "A single bomb of this type exploded in a port might very well destroy the whole port along with the surrounding territory."

Roosevelt was persuaded, and the Manhattan Project began. After sending the letter, Einstein had nothing to do with the bomb project. He learned that a bomb had been made only after Hiroshima was incinerated in 1945. Immediately, his pacifist desires resurfaced, and Einstein joined the many other scientists calling for no further use of the weapon. He urged the creation of a world government, saying "we must not be merely willing, but actively eager to submit ourselves to the binding authority necessary for world security."

In 1950 Einstein published yet another try at a unified field theory but it too was respectfully disregarded. Shortly afterward his health began to deteriorate seriously. He could no longer play the violin or sail his boat. In 1952 he was invited to become president of the fledgling state of Israel but declined. "Politics," he had once declared, "are for the moment. An equation is for eternity." In 1955 Albert Einstein died in a hospital bed at Princeton.

See ATOM THEORY; RELATIVITY; QUANTUM THEORY; UNIFIED FIELD THEORY; PLANCK; BOHR.

ELDREDGE, NILES, 1943–   , American evolutionary biologist. See PUNCTUATED EQUILIBRIA; MODERN SYNTHESIS.

ELECTRICITY—Electrons are subatomic particles that always carry a negative charge and which readily travel from one atom to another. Electrons either moving or stationary account for the phenomenon called electricity. Electrons tend to flow from a region with a surplus of loose electrons (negative charge) to a region with relatively fewer loose electrons (positive charge).

Friction and other forms of energy can cause a surplus of electrons to accumulate in some object, giving it a strong negative charge. Since opposite charges attract, an object with many free electrons will be strongly attracted to another object with relatively fewer. If the imbalance is great enough, the electrons will jump the gap between the objects, causing a spark which may be small (between a finger and a doorknob, for example) or large (lightning). This is static electricity.

An electrical current, by comparison, involves a flow of electrons through a wire or other conductor. This flow can be induced by moving a magnet near a wire, since electricity and magnetism are two forms of the same basic phenomenon. All electric generators, or dynamos, are devices for efficiently moving magnets relative to wires. When a heavy enough flow of electrons passes through certain metals, they heat up—either just enough to toast bread or enough to glow white hot, as in a light bulb.

See ELECTRON; CELL (ELECTRICAL); CAVENDISH; FARADAY; MAXWELL (all of whom made key discoveries about the nature of electricity).

ELECTROMAGNETISM—The phenomenon responsible for light and other electromagnetic waves, such as radio waves and x-rays, electromagnetism is one of the four fundamental forces of nature (the others are gravity and the two nuclear forces called the strong and weak forces).

The force of electromagnetism binds oppositely charged particles to one another. This effect keeps atoms together, binding electrons (carrying a negative charge) to atomic nuclei (which contain protons with positive charges). Electromagnetism is also what binds atoms into complex molecules.

When charged particles such as electrons move, they produce an electromagnetic field around the path through which they move. The field reaches out indefinitely in all directions, though it becomes more diffuse with distance. Depending on the characteristics of the particles' movement, the field may be of the sort used by radio and TV stations to carry programs. When a radio or television station broadcasts, it varies the flow of electrons into an antenna in such a way that the variations encode corresponding variations in sound or picture. The electromagnetic waves emanating from the antenna vary accordingly. Under other arrangements, the field may consist of x-

The Electromagnetic Spectrum

All the forms of radiation in the chart at the right represent different names and uses for the same thing —electromagnetic waves. They differ only in frequency and wavelength but these differences endow a given portion of the spectrum with special abilities.

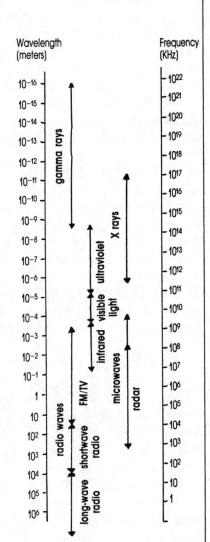

**low frequency,
long wavelength**

**high frequency,
short wavelength**

rays, or, if the wavelength happens to stimulate nerve impulses when it enters the eye, it will be visible light.

One of science's deepest mysteries is the nature of electromagnetism itself. Sometimes its effects are best explained by thinking of the electromagnetic force as carried by particles called photons. Electrons and protons bind to one another by continually exchanging photons. Radio and light are showers of photons flying out from a source. At other times, electromagnetism is better explained as a wave—not a substance in itself but a disturbance in another substance. Neither explanation is adequate to describe all the properties of electromagnetism.

See FORCE; QUANTUM THEORY (which holds light comes in packets, or quanta, called photons); FARADAY; HELMHOLTZ (both of whom helped show that electricity and magnetism were alternative manifestations of a single force).

ELECTRON—These are the particles that orbit the nucleus of every atom, each element having a different number of them. Hydrogen atoms, for example, have one electron. Oxygen has eight. Iron has twenty-six. Gold has seventy-nine. In the stable form of each atom the number of electrons is balanced by the number of protons in the nucleus. The balance is not one of weight, or mass (protons are 1,836 times heavier than electrons), but of electrical charge. Each electron has one unit of negative electrical charge and each proton has one unit of positive charge.

When two atoms form a chemical bond, they may do so in two ways. One may donate an electron to the other, forming a relatively weak bond called an ionic bond. Or the two may share electrons, with the shared electrons actually orbiting both nuclei. This is a very strong attachment called a covalent bond. The discovery of the nature of chemical bonding was one of the major advances of science.

Electrons are also the things that flow through a wire as electricity. They can even be made to jump off a wire and fly through a vacuum, as happens with the "electron gun" inside a TV picture tube. When the electrons hit a point on the phosphor-coated screen of the tube, a tiny dot glows. Since electrons are charged, however, their path inside the picture tube can be deflected by a magnetic field. In fact, a constantly changing magnetic field makes the beam of electrons in the tube scan the entire picture surface of the tube, planting dots that fade just before the next scan strikes that spot on the tube a fraction of a second later.

The nature of the electron, beyond its mass and charge, remains a great mystery. Whereas subatomic particles such as protons and neutrons are now believed to be made of smaller particles called quarks, electrons do not seem to be divisible. They may be fundamental particles just like quarks. As such, an electron is probably most easily conceived of as a tiny bundle of energy which, because of the way it is bundled, amounts to a particle of matter.

See ATOM STRUCTURE; CHEMICAL BONDING; THOMSON, J. J. (the discoverer of the electron); QUARK.

ELEMENT—Each atom in the universe represents one kind of element. There are some 105 kinds of elements known. Of these, 92 occur in nature and the rest have been manufactured by scientists using particle accelerators. All the millions of chemical compounds are made of combinations of atoms of the 105 known elements in various numbers and proportions. Water, for example, is a compound. One molecule of water is made of three atoms: two atoms of the element hydrogen and one atom of the element oxygen, resulting in the formula H_2O.

The most abundant element in the universe is hydrogen, which accounts for about 90 percent of all atoms. Helium, the next most abundant, accounts for another 9 percent. Thus, most of the elements we take for granted on Earth are relatively rare ingredients of the universe. Of the elements in the Earth's crust, oxygen is the most common, amounting to 47 percent of the weight of the crust. Silicon is second at 28 percent. Aluminum is the most common metallic element, followed by iron, calcium, sodium, potassium, and magnesium. These eight elements—oxygen through magnesium—make up about 99 percent of the Earth's crust. The most abundant elements in living organisms are carbon, hydrogen, oxygen, and nitrogen. These four elements can be combined in different patterns and ratios to create the thousands of different kinds of protein molecules that make up a living cell.

See ATOM; MOLECULE; ISOTOPE; PERIODIC TABLE; MENDELEYEV; DALTON.

ELEMENTARY PARTICLE—Also called subatomic particles, these come in two kinds: the particles of which atoms are made—protons, neutrons, and electrons—and a host of other particles emitted by atoms as they are "smashed" in accelerators or, in nature, hit by flying subatomic particles. The emitted particles—various kinds of mesons, neutrinos, and many others—do not exist as such within the atom, but form in the process of the smashing and then disintegrate, usually within a few millionths of a second. The only stable particles created in the smashing of an atom are neutrinos. Yet another kind of elementary particle is the photon, the particle of light that is also a wave. Under certain circumstances atoms also emit photons.

The reason for this odd state of affairs—the creation of new particles when an atom is hit hard—is that two of the basic elementary particles, the proton and the neutron, are actually *not* fundamental particles. They are made of even smaller particles called quarks, of which there are six different kinds. Depending on how these quarks split up or recombine during the moment of smashing of an atom, they form one or more of the short-lived particles. Since the combination of quarks is unstable, it quickly breaks up. The only stable combinations of quarks are the ones that make protons and neutrons, but even neutrons are not so stable. An isolated neutron has a half life of only about fifteen minutes, but when bound in an atom's nucleus it is stable.

The big question facing particle physicists now is whether quarks are in fact truly fundamental particles. Will it someday become possible to crack open quarks and find they are made of still smaller particles?

See QUARK; ATOMIC STRUCTURE; QUANTUM THEORY; BOHR; RUTHERFORD.

EMPIRICAL/EMPIRICISM—Philosophers of science can get pretty sticky about the meaning of empiricism, but in the way this term is usually used it simply refers to an intellectual approach in which concrete events and experiences are taken as the source of knowledge, in preference to pure intuition or supposition. If you have a problem, the empirical approach is to do an experiment or make an observation to see what's what.

ENERGY—There are many ways to define energy. The one closest to everyday experience is that it is the capacity to do work, usually in the sense of moving an object, whether a person, a piston, or a proton. Any form of energy can be converted to any other. For example, the chemical energy locked in food molecules can be released by living cells to move muscles, and the chemical energy locked in petroleum molecules can be released by burning to move vehicles. In the large sense, energy is never used up; it is simply converted from one form to another, but often with some of the energy becoming heat that dissipates and cannot be recaptured.

Fundamentally, there are only two things in the universe—energy and matter. Einstein's relativity theory shows that energy and matter are, in fact, alternative forms of the same single fundamental thing.

See CONSERVATION OF ENERGY; FORCE; HELMHOLTZ; EINSTEIN; RELATIVITY.

ENTROPY—Entropy is a measure of the disorder in a closed system, and the law of entropy says that whenever a change occurs in that system, the amount of entropy (disorder) increases. Ultimately, the law of entropy says, everything that is organized will break down or run down unless it is maintained. Examples of entropy's workings can be found throughout the everyday world. Desks will get messy. Cars will wear out. Stars will blow up. Without librarians, books get scattered and jumbled. Lacking nutrients, organisms die and rot. In each case, a highly organized system will inevitably proceed to a state of disorder and chaos unless energy (which is equivalent to work) is brought into the system to re-establish order. Fortunately, the law of entropy does permit order to increase in open systems. These are regions into which it is possible to import energy from an outside source.

The law of entropy is actually a principle of thermodynamics. It is, in fact, one way of expressing the second law of thermodynamics, which can be stated to the effect that heat will not spontaneously flow from a colder body into a hotter body. The concept of entropy is statistical and perhaps best illustrated through the following example. Take a bottle and fill it one-quarter full with salt. Then pour pepper on top of the salt until the bottle is half full. Cap the bottle, creating a closed system. At this point the system is highly organized—all of the black pepper is in one place atop all of the white salt. Now anything that happens to the system can only decrease its order. Shake the bottle and the salt and pepper begin to mix; the disorder is increased. Every time you shake the bottle the disorder grows, until the two substances are completely mixed. The shaking was a random event, but no amount of shaking is going to result in the highly improbable event of the salt and pepper resegregating themselves. The only way to increase the order is to open the system and bring in some outside energy. It could be done, say, by sorting the particles with tweezers and a magnifying glass.

A major open system in which the law of entropy does not prevail is the realm in which life exists. Sometimes people claim that life is a violation of the second law of thermodynamics (or that evolution is a violation of this law), without realizing that the biosphere is not a closed system. It is continually bathed in massive amounts of solar energy that plants capture and pass on to the rest of us.

See THERMODYNAMICS.

ENZYME—An enzyme is a molecule that causes other molecules to undergo specific chemical reactions, breaking them into smaller pieces or making them combine with other substances. Enzymes are a form of catalyst. Enzymes that break down other molecules are responsible for digestion. All enzymes are special forms of protein. Living cells may produce hundreds or thousands of different kinds of enzymes in order to carry out various metabolic processes. Synthetic enzymes, made in a factory in imitation of natural enzymes, are often added to laundry detergents to help break down the molecules responsible for food stains.

ERATOSTHENES, 276 B.C.–195 B.C., Alexandrian mathematician, geographer—Eratosthenes, equipped with nothing more than a good mind and a grasp of geometry, calculated the circumference of the Earth some 1,700 years before anyone traveled around it. His figure was within 0.2 percent of the true circumference.

Eratosthenes was born in Cyrene, which is now in Libya, and educated in Athens. He returned to North Africa and spent most of his professional life as librarian at the fabled 700,000-volume library of Alexandria. This institution was a center of learning and science that would not be surpassed until the Renaissance. Unfortunately, it was destroyed in its prime when Julius Caesar's conquering army burned it, an act that halted intellectual progress for many centuries.

Besides being a geographer and mathematician, Eratosthenes was also an athlete of some standing. He was known as a "pentathlus," or champion of the five sports of the Greek pentathalon—which at that time included running, jumping, javelin throwing, wrestling, and quoit throwing. Eratosthenes was also known as Beta, the second letter of the Greek alphabet, signifying his stature as second only to Plato.

Eratosthenes produced the first maps of the world that used the system of latitudes and longitudes. As part of his effort to map the Earth, he calculated how big it was. He did this by noticing that at noon on a certain day the sun was directly overhead (at its zenith) at the town of Syene (now Aswan, Egypt). He knew this because only at noon on that day did it shine straight down to the bottom of a deep well in the town. Eratosthenes found that on the same day at Alexandria, which was to the north of Syene, the sun was south of its zenith by an angle of 7 degrees, 12 minutes. He determined this from the length of a shadow cast by a vertical stick. This angle is one-fiftieth of a circle, and Eratosthenes therefore reasoned that the distance between Alexandria and Syene must be one-fiftieth of the circumference of the Earth.

To make this method work, Eratosthenes had to assume that the sun was extremely far from the Earth—so far that the rays reaching Earth were all traveling virtually parallel to one another. He was, of course, right. In fact, his sense of the position of the Earth in relation to the sun was accurate enough that he was able to improve the measurement of how much the Earth tilted on its axis, coming quite close to the modern figure.

ETHER—Not the anesthetic, but the ethereal substance that physicists once believed had to permeate the whole universe so as to carry light. Light was believed to be a wave, and like all waves, to consist of a disturbance in some substance. Ether was the name given this substance, which no one could detect. The Michelson-Morley experiment to detect the motion of the Earth through the ether forced scientists to the conclusion that the substance did not exist.

See LIGHT, NATURE OF; RELATIVITY (which discusses the Michelson-Morley experiment).

EUCLID, c. 330 B.C.–c. 260 B.C., Greek mathematician— Euclid didn't invent plane geometry as is widely believed, but he did take what was known in his time and re-

reorganize it into such a logical and coherent presentation, entitled "Elements," that his name has become virtually synonymous with geometry. Euclid also established the form of geometric proof so familiar to high school students. In fact, euclidian proofs became so persuasive as devices for waging mathematical arguments that even Newton used them in his classic *Principia Mathematica*. Had Newton instead used the calculus—which he invented—for waging such arguments, the *Principia*, his scientific masterpiece, would be more comprehensible to today's mathematicians.

Virtually nothing is known of Euclid's personal life except that he founded a school in Alexandria, Egypt—the great scientific and mathematical center of the ancient world. Even his birth and death dates are guesses.

EVOLUTION—Why do whales have useless thigh bones buried in the muscles of their hindquarters? Why is the genetic code in the chromosomes of a human being the same as the genetic code in every other living creature on earth? (Biochemically, we know that lots of other codes would work just as well.) Why have human bones never been found in the rock layers that contain dinosaur bones?

The answers to these questions, and countless others that have to do with life on Earth, can be found and understood through the study of the process called organic evolution. This is the concept that all forms of life are descended from one common ancestor that lived hundreds of millions of years ago. The process starts with the assumption that some of the offspring of that original species— perhaps a creature like today's bacteria—differed slightly from their parents. If the differences allowed these offspring to live under conditions that their parents found intolerable (warmer or colder water, for example), the offspring could exploit a virgin environment without competition from their unchanged relatives. With time and the passage of many generations, organisms might accumulate a wide variety of differences from their forebears.

After some 700 million years of this process—thousands

of new varieties branching off and finding themselves
adapted to new environments (also called ecological
niches)—the world became filled with worms, lobsters, insects, fish, amphibians, reptiles, birds, and mammals. And
also with a succession of plants from algae to mosses,
ferns, pine trees, and all the flowering plants.

Charles Darwin wasn't the first to think of evolution;
the ancient Greeks, including a man named Anaximander,
had the idea. But they did not make a good case for how it
happened. Darwin, after his five-year voyage around the
world, proposed the mechanism of natural selection, which
others called "survival of the fittest." Darwin's ideas were
based on several observations. First, that animals have
more offspring than can survive in their environment. An
insect pair, for example, may produce thousands of offspring, but only two need to survive to keep the population stable. Second, that there is competition among the
offspring. Thus, those with any built-in advantages will be
more likely to survive, and therefore to have offspring
that will inherit the advantages.

Offspring get their advantages as the result of a totally
random change in their genes—a rearrangement of existing genes or a mutation that damages a gene in some fortuitous way. Most genetic changes put nonsense into the
genetic message, and the individual in which the change
occurs dies. But once in a while a change will be useful,
and nature will select that individual to prosper. It was
this that Darwin called natural selection. It simply
amounts to nature, or the environment, doing what a
farmer does when he selects prize cows to breed and culls
out the runts.

Darwin thought the process of natural selection worked
gradually, with a new variety slowly becoming more and
more different from its ancestor until it was eventually so
different that we would call it a different species. Today
some evolutionary biologists think that the biggest effects
of evolution occur rapidly. They believe that when a new
species is splitting off from an old one, it evolves very
rapidly. Then, they say, the new species virtually stops

evolving and remains almost unchanged for the rest of its existence. This new modification of the theory of evolution is called punctuated equilibrium. According to it, most species are in an evolutionary equilibrium that is occasionally punctuated by a rapid burst of evolution when a new species splits off. The originators of this new idea are Stephen Jay Gould of Harvard University and Niles Eldredge of the American Museum of Natural History in New York.

See DARWIN; MODERN SYNTHESIS (which established the genetic basis of evolution); PUNCTUATED EQUILIBRIA.

EXON—A portion of a DNA sequence that contains properly coded genetic instructions. The term is used to contrast this part of a DNA sequence, of genes, with intervening sequences of nonsense DNA called introns.

See NONSENSE DNA; GENES, HOW THEY WORK; GENETIC CODE.

EXPERIMENT—One of the chief ways science advances is by testing new ideas in an experiment. An experiment is any procedure in which a presumed cause is tested to see whether it produces the effect that a hypothesis predicts. Experimental results, however, are meaningless unless they can be compared with results observed in controls. These are experimental subjects (chemicals, animals, people, or whatever) treated exactly like the experimental subjects except that they are not exposed to the presumed cause. If the effect shows up in the controls, it is clear that the hypothesis (linking the presumed cause and the effect) is wrong.

See PART 1: WHAT IS SCIENCE?

EXPRESSION—Usually this term refers to the process in which the genetic instructions encoded in DNA are carried out. Genes may be inactive for some period, during which gene expression is not going on. The process of gene expression involves transcribing the gene's coded message from DNA (which never leaves the nucleus) into ribonucleic acid, or RNA, which exits the nucleus, carrying the code to the cell's protein-making machines, the ribosomes. The ribosomes "read" the message carried by the RNA and assemble the specified amino acids to make the proper protein molecule.

See GENES, HOW THEY WORK; GENETIC CODE.

FARADAY, MICHAEL, 1791–1867, English chemist and physicist—Michael Faraday, often regarded as one of the world's greatest experimental scientists, made many of the fundamental discoveries and interpretations that revealed magnetism and electricity to be two manifestations of the same physical force. He discovered that electricity could be used to make objects move (the principle underlying the electric motor), and that moving a magnet relative to a wire would create electricity in the wire (the principle underlying the electric generator). Faraday's ideas, made mathematically precise and more broadly applicable by James Clerk Maxwell, a Scottish physicist, underlie all modern uses of electricity and electronics. Yet Faraday had no formal education beyond grammar school.

Faraday is also remembered as an eager and highly successful popularizer of science. In 1826 he started the famous Christmas lectures for children, given annually at the Royal Institution in London, where he worked all his life. Faraday gave nineteen of the lectures, and his skill as an expositor of science is credited with awakening the scientific interests of many young people.

Faraday's own scientific interest arose even more informally. As the son of an impoverished blacksmith, he had to leave school at the age of thirteen to take a job working for a bookbinder. Among the binder's clients was the *Encyclopaedia Brittanica* and, so the story goes, Faraday discovered his interest in science when he read the 127-page entry on electricity. This stimulated him to buy some simple equipment and try his own electrical experiments.

At the age of twenty-one he happened to attend a series

of lectures given at the Royal Institution by Humphry Davy, and resolved to make science a career. (The experience would prompt him later to establish the Christmas lectures for children.) Faraday went back to his bookbinding shop, bound the detailed notes he had made of Davy's lectures, and sent the volume to Davy, hoping to impress him enough to get a job. Luck was with Faraday, for Davy soon was temporarily blinded during an explosion in his laboratory, and needed a helper to carry on with his work. Faraday got the job, but at a salary less than he was making as a bookbinder's apprentice. In twelve years, however, he had worked his way up to being director of the laboratory, and a few years later to an endowed professorship.

Still, the pay was meager, and Faraday had to supplement his income with consultancy fees and a part-time lectureship elsewhere. At the height of his electrical experiments, however, Faraday gave up his outside jobs to concentrate on his science. His ensuing financial difficulty prompted friends to try to arrange a government pension. Faraday had to put his request to the prime minister, who promptly sneered at the idea of pensions, but apparently would grant one. Faraday's pride was so hurt by the sneer, however, that he refused the pension. He also refused other awards, turned down a knighthood, and twice declined to accept the presidency of the Royal Society, the most prestigious scientific society in Britain. "I have always felt," Faraday wrote, "that there is something degrading in offering rewards for intellectual exertion, and that societies or academies, or even kings and emperors should mingle in the matter does not remove the degradation."

Faraday's major electromagnetic discoveries were prompted by an observation in 1820 by Hans Christian Oersted, a Danish physicist. Oersted found that if he moved a wire carrying an electrical current (from a battery) near a magnetic compass, the compass needle would be deflected. The direction of the deflection depended on which way the current was flowing in the wire. The experiment, which many others immediately repeated and pursued to new interpretations, marked the beginning of the modern understanding not only of electricity and magnetism but of all the fundamental forces of the universe.

Faraday confirmed Oersted's findings and greatly extended them. First, he had to reject the prevailing notion that electricity was a fluid. No fluid can produce an effect beyond the boundaries of its stream, whereas electricity clearly sends out some kind of force into the air around the wire through which it is flowing. Faraday's genius was in perceiving what he called the force field surrounding the wire. Somehow the wire's electrical field could interact with the field of a magnetized compass needle, making the magnet move. Faraday quickly tried the opposite experiment—moving a magnet near the wire. He discovered that this caused electricity to flow in the wire, as long as the magnet and the wire were kept moving with respect to one another. But while it did not matter which was moving, the wire or the magnet, it did matter which way the wire was held with respect to the magnet. Depending upon how the wire was held with regard to the magnet, the amount of current produced could vary considerably. From these observations, Faraday inferred the existence of "lines of force" emanating from the magnet and the wire, following curving paths. When the wire cut these invisible lines, an electric current flowed through it. This is the principle underlying the dynamo, or electrical generator.

Oersted's discovery and Faraday's further work established that electricity and magnetism were alternative forms of the same fundamental kind of energy. This led to one of Faraday's strongest convictions: "that the forms under which the forces of matter are made manifest have one common origin." This was the beginning of what has come to be known as the unified field theory, which Einstein and others have pursued for many years, especially in the hope of showing that gravity and electromagnetism have a common basis, a supposition that Faraday himself held. In more recent times Faraday's hunch has been partially vindicated with the demonstration that electromag-

netism and the weak force are alternative forms of the
same force, this now called the electroweak force. Faraday also did another series of experiments to show that all
the forms of electricity then known were the same. In
Faraday's time people thought there were five different
kinds of electricity—static electricity (made by friction,
such as that between a glass rod and fur), voltaic electricity (made by a battery), magneto-electricity (made by
a magnet moving relative to a wire), thermo-electricity
(made by heating the junction of wires made of dissimilar
metals), and animal electricity (made by animals like electric eels). Each, he showed, would magnetize iron, make
sparks, and generally produce the same effects.

After several years of intensive work on his electromagnetic theories, Faraday suffered a severe mental breakdown and retired to a country house for some three years.
It is not clear what the problem was but Faraday refused
to see visitors during the time and wrote that he suffered
"ill health connected with my head." In 1844 he emerged
from seclusion and worked again, this time searching for a
unity between light and electromagnetism. He found a relationship by showing that when polarized light is passed
through a magnetic field, the plane of polarization is rotated. It was a discovery comparable to Oersted's and
Maxwell would take it further, strengthening the idea
that light is a form of electromagnetism.

Gradually Faraday's scientific powers waned along with
his health, and at the age of sixty-seven he retired. He
died nine years later.

See ELECTROMAGNETISM; MAXWELL; DAVY; UNIFIED
FIELD THEORY.

FERMENTATION—The process in which bacteria and
yeast organisms feed upon complex organic matter and
break it down into simpler molecules. For example, the
yeasts that grow in grape juice use an enzyme to digest
the sugar called glucose into two simpler substances—
ethyl alcohol (the drinking kind) and carbon dioxide. The
breakdown releases some energy, which is what the yeast

is after. Fermentation is an ancient process used in making many alcoholic beverages, bread, yogurt, tofu, and many other foods. Until the nineteenth century most people considered the process almost magical. Then Louis Pasteur showed that it was, in fact, a result of the action of microscopic living organisms.

See PASTEUR.

FITNESS—Not the meaning used by runners but by evolutionary theorists. Fitness is a measure of the reproductive success of any individual of any species. The more viable offspring an individual produces, the more individuals there will be in the next generation bearing the same genetic traits. Thus, "survival of the fittest" does not necessarily mean survival of the strongest or fiercest; an individual animal that is better able to hide from its enemies (and therefore live longer to produce more offspring) may, for example, be more fit than a species mate ferocious enough to get into lots of battles but whose ferocity cuts short its life. Any trait that lengthens an individual's lifespan, improves its access to members of the opposite sex, or both will enhance fitness. This is the fundamental process by which natural selection works.

Degrees of fitness are ordinarily comparable only within a species. Sometimes, however, the anatomical or behavioral differences that underlie differences in fitness can make one population of individuals so different from other populations in the same species that the two populations no longer interbreed. As a result, this new and distinctive population may evolve into a new species.

See EVOLUTION THEORY; DARWIN; NATURAL SELECTION.

FLEMING, ALEXANDER, 1881–1955, British bacteriologist—Alexander Fleming discovered penicillin, the first antibiotic, in 1928, but assumed that it was no good for fighting infection and never pursued his discovery. Not until twelve years later did an Oxford University team "rediscover" the drug and show that it had an almost miraculous ability to cure many of the most serious infections known.

Nevertheless, Alexander Fleming got the Nobel Prize and just about all the glory for this work. His is one of the great stories of scientific discovery—the finding that a humble strain of bread mold manufactured a substance that killed bacteria. Howard Florey and Ernst Chain, the Oxford team that showed penicillin could cure human disease and end suffering on a massive scale, remain virtually unknown.

Fleming's discovery was an accident. He was a bacteriologist and was working with the bacteria known as staphylococcus that cause the familiar "staph" infection. He inadvertently left a dish of the bacteria uncovered and returned to find several spots on it where there were no bacteria. In the middle of these clear spots, Fleming found small, growing colonies of the mold species called *Penicillium notatum*. Fleming surmised that the mold must have been exuding some substance that killed the bacteria, but he was unable to identify the substance, although he went ahead anyway and gave it the name penicillin. Only after World War II broke out and the need for weapons against infection became severe was the substance isolated.

To some historians of science, the case of Fleming illustrates the difference between a great scientist and a scientist who makes a great discovery. Had Fleming been truly great, historians have claimed, he would have pursued his discovery instead of letting it lapse.

FLOREY, HOWARD, 1898–1968, Australia-born British pathologist. See FLEMING.

FOOD CHAIN—The path by which energy is relayed from the sun to plants to plant-eating animals and then to meat-eating animals. Actually the natural world is a food web, a tangle of interlocking food chains in which not only energy is passed along, but various chemical compounds synthesized in the body of each link in each chain. The food web is also circular, because once the animal at the top—the ultimate consumer such as a lion or human or hawk—dies, many smaller organisms feed from its decomposing corpse. If dung beetles practiced ecology, they would no doubt draw food chains showing dung beetles at the top and ignoring the birds that eat dung beetles.

FORCE—The term is used in two related ways in science: to refer to the fundamental interaction between particles of matter that bind these particles together or propel them apart, and to refer to everyday phenomena that result from fundamental forces acting in concert.

The more familiar form of force is the latter one, such as the phenomenon that makes an object change its motion. The force of gasoline burning in the combustion chamber of an automobile motor pushes the piston, which makes the crankshaft rotate. The force is further transmitted from the crankshaft to the car's wheels, causing the car to move. The force of warm air rising can create winds that blow down trees. The force of muscles contracting can make balls hurtle through the air. In all these cases, force is a mechanical agent. It is merely transferred from one body to another. The energy consumed as a force acts may come from burning fuel or, in a living organism, burning food, or something else, such as an impact from another moving object. Once the force has acted on an object, it may endow that object with either of two other forms of energy: kinetic energy or potential energy. Kinetic energy is the energy of motion. Should the moving object stop, it will give up its kinetic energy, either by transferring it to another object that the first object hits and which will move accordingly, or by converting the kinetic energy to heat energy. Potential energy is the energy of position or, rather, of being in a special position. An object on a shelf, for example, has potential energy corresponding to the force needed to lift it that high. As long as the force of gravity pulls on the object, it contains potential energy that may be converted to kinetic energy if the shelf breaks.

The deeper use of the term force has to do with the fundamental nature of all existence—the irreducible phenomena, or interactions, that underlie all forms of energy and matter. In this sense, there are only four kinds of forces at work in the universe. The most familiar is gravity, or gravitation, which holds the planets, stars, and galaxies together. Less familiar is electromagnetism, which holds electrons in orbit around atomic nuclei. Electromagnetism also underlies electricity, light, radio and television broadcasting, and magnetism. The other two forces in the universe act within the atomic nucleus. One, the so-called strong force, binds protons and neutrons together to form atomic nuclei. The weak force governs radioactive decay and, inside stars, causes thermonuclear reactions to emit particles of light called photons.

All forms of energy can be accounted for by the four forces. For example, the energy that comes from burning fuel or food actually comes from the liberated electromagnetic force that bound several atoms of the fuel material into one molecule. Animal cells continually break apart food molecules and recombine them with oxygen, a process that transfers the electromagnetic force released in the breaking apart from one place to another within the cell until, as a final step, a muscle fiber contracts, producing a mechanical force.

See GRAVITY; ELECTROMAGNETISM; STRONG FORCE; WEAK FORCE.

FOSSIL—Any trace of a living organism from the past counts as a fossil; it is not simply one or more ancient bones. The hominid footprints made more than 3.5 million years ago in Tanzania are fossils. So are the burrows left by worms hundreds of millions of years ago. Even the cut marks left on butchered animal bones of a million years ago are sometimes spoken of as fossils of behavior. Nor are fossil bones merely stone replicas of the original. Mineralization does not so much replace the original bone as infiltrate it with atoms of stone, and most fossil bones contain much of the original material of the bone, including the calcium and the bone protein called collagen. Even dinosaur bones still contain the same stuff that was once alive.

See CUVIER (the first scientist to take fossil bones seriously); LYELL (the geologist whose theories showed fossils to be very old); DARWIN.

FOX, SIDNEY, 1912– , American molecular biologist. See LIFE, ORIGIN OF.

FRANKLIN, ROSALIND, 1920–1958, British molecular biologist, crystallographer. See DOUBLE HELIX.

GALAXY—A grouping of billions of stars and vast clouds of dust orbiting a common center. Earth is on the outskirts of one galaxy called the Milky Way, which contains an estimated 800 billion stars. There are an estimated 200 billion other galaxies in the universe, though only three can be seen from the Earth with the naked eye. (They look like fuzzy blobs, an appearance that produced the old name of nebulae for these objects.) Telescopes can, however, spot thousands of galaxies, ranging in shape from globular clusters to elliptical clusters to spirals like our Milky Way. The name Milky Way derives from the ancient Greeks, who saw the luminous band of light arcing across the night sky and called it a "milky circle." In fact, the word *galaxia* is Greek for milky (for which reason it is technically redundant to speak of the Milky Way galaxy). The Romans called it the *via lactea*, or milky way. With the invention of telescopes, it became clear that the luminous arc was, in fact, made of individual dots of light, the stars. The swath of the Milky Way represents an edge-on view of our galaxy.

Galaxies are among the more enduring features of the universe. Within them generations of stars are born, live, and die. As the universe expands, the galaxies are rushing away from one another (the expansion does not take place within a galaxy). Still, galaxies are often organized into clusters. The Milky Way is in a cluster of galaxies that astronomers call the Local Group.

See BIG BANG; STAR.

GALILEO GALILEI, 1564–1642, Italian astronomer and physicist—Galileo, a college dropout, was the first person to make and use telescopes that could be used for studying celestial objects. His observations led him to confirm the theory of Copernicus, put forth almost a century earlier, that the Earth orbits the sun. Because his advocacy of this position conflicted with Church dogma, Galileo was tried by the Holy Inquisition, forced to recant his beliefs (officially but not in his own mind), and held under house arrest for the last eight years of his life.

Less well known are some of Galileo's other discoveries, all made using his 32-power telescope: that the Moon has mountains, that the Milky Way is made of separate stars, that Jupiter has moons, that Saturn has rings, and that the sun has spots.

Galileo was born at Pisa in 1564, the son of a musician. The story is told that one day, after entering the University of Pisa as a medical student, Galileo was in the cathedral and happened to notice that a hanging lamp was swinging. The service must not have been too compelling for the young Galileo, for he timed the lamp's movements. He found that it always took the same time to complete an arc, even as the swinging subsided into shorter and shorter arcs. The observation turned Galileo's mind away from medicine, which had never interested him much anyway, and toward mathematics as a career. More specifically, the incident awakened Galileo to the intimate linkage between physical phenomena and mathematics.

Before he had earned his medical degree, however, Galileo had to drop out because his family lacked the money to keep him in school. Nonetheless, he found work as a private tutor and as a lecturer at the Florentine academy and, at the age of twenty-two, published his invention of the hydrostatic balance, a device for measuring the pressure in a fluid. The invention made his name known throughout Italy. Other papers followed, and Galileo was soon hired as a lecturer at the University of Pisa. Three years later he moved to the mathematics chair at Padua, a better-paying job that Galileo needed to make ends meet.

The young professor, however, never got on well with his fellow faculty members because he had already formulated the rebellious intellectual stance that marked him for greatness and, of course, for the Inquisition. Even before he made his first telescope, Galileo was already challenging the scientific orthodoxy of his day, which was based almost entirely on Aristotle. Thus, for example, while Aristotle taught that heavier objects fall faster than light objects, Galileo's experiments led him to reject Aristotle and to conclude that all objects fall at the same rate. (Wind resistance complicates the observation of this because its effect is proportionately greater on objects that are very light or that have a large surface area. It will be negligible, however, if the experiment compares objects such as a bowling ball and a marble.) The legend that Galileo proved his point by dropping weights from the leaning tower of Pisa seems not to be true. Some historians say that he merely dared his detractors to try such a public experiment.

Galileo did not keep his unorthodox views to himself. He campaigned vigorously for them, much to the resentment of his colleagues. He did, however, keep one opinion to himself. This was his conviction that Copernicus was right about the Earth orbiting the sun. The prevailing view, named for Ptolemy, claimed that all heavenly bodies rotated about a stationary Earth. Without experimental evidence to the contrary, Galileo thought it better to keep quiet.

Then, in 1609, at the age of forty-five, Galileo heard about the invention of the telescope by Lippershey. Within the same year, Galileo built one for himself. It was weak by modern standards, with only about a threefold magnifying power. But he quickly devised methods of improving it, and made better ones that could magnify thirty-two times. Then, still within the same year, Galileo made a series of major astronomical discoveries—from the mountains of the moon to the existence of moons orbiting about Jupiter. An approving Venetian senate granted Galileo a lifetime appointment at Padua. Within months, however, Galileo left Padua to work for the duke of Tuscany as his official "philosopher and mathematician." The position finally gave Galileo enough income to do less teaching and devote more time to research.

It was at about this time that Galileo began to run afoul of the Church. He went to Rome and demonstrated his telescope for the pontifical authorities. They were, at first, pleasantly amazed. Encouraged, Galileo decided to go public with some of his interpretations. He had seen that the moons of Jupiter revolved about the planet. He had seen that sunspots move across the face of the sun, suggesting that the sun was spinning. He put it all together and concluded that Copernicus was right.

Had Galileo done this in some dull, academic paper, he might have stayed out of trouble. As it happened, though, he was an excellent prose stylist and his papers won wide readership beyond the scientific community. When the Aristotelian professors saw Galileo, their enemy, winning popular support, they united against him and reminded Church leaders that Copernicanism conflicted with a literal interpretation of the Bible. The Church, then much embattled against the Protestant reformation, feared losing popular support on yet another front. Churchly fundamentalists denounced Galileo to the Inquisition.

Hoping to forestall a formal attack, Galileo wrote letters reminding Church authorities of the practice, already established, of interpreting the Bible allegorically whenever it conflicted with scientific findings. He did not succeed. The Church's chief theologian clung to the orthodox belief that mathematical hypotheses are unrelated to physical phenomena. A church decree was promulgated, forbidding Galileo from believing or defending his views, although they could still be discussed as a "mathematical supposition."

Galileo complied and for seven years stayed in virtual seclusion at his home near Florence. Then, when a new paper on comets attacked Galileo, he felt compelled to reply. This took the form of a book expounding his method of arriving at physical truth through observation, the use of mathematics to make predictions, and the use of experiments to confirm those predictions. It was the first important statement of a method that has since become a bulwark of science. "The book of nature," Galileo said in one of his most famous passages, "is written in mathematical characters." Galileo dedicated his book to the new pope, Urban VII, who had been a friend and protector.

Urban replied warmly, and Galileo felt encouraged to go to Rome and ask for a revocation of the decree restricting him. The pope refused, but did allow Galileo to write about Copernicanism as long as he also gave equal time to the Ptolemaic system and as long as he came to the prescribed conclusion that nobody could presume to know how God made the world.

Galileo then spent several years writing just such a book. It came out in 1632 with the full approval of the Church, and immediately won international acclaim as a literary and philosophical masterpiece. Only later did the pope learn that although the book stuck to the letter of the pope's order, the eloquence and power of the argument for a Copernican universe far outweighed the meek conclusion that nobody could presume to know how God made the world.

Angered, the pope ordered Galileo prosecuted. At the age of sixty-nine and despite pleas of illness, Galileo was brought to trial in Rome. A document was produced, claiming that he had been specifically ordered during his first tangle with the Church not to discuss Copernicanism "in any way." Historians would later conclude that the document was a forgery. Although Galileo said he never saw the document, he was convicted. The sentence called for Galileo to "abjure, curse, and detest" his errors, and for his imprisonment. The pope commuted this sentence to house arrest for life, and Galileo died eight years later, a prisoner in his own house.

See COPERNICUS; COPERNICANISM.

GALVANI, LUIGI, 1737–1798, Italian biologist. See VOLTA.

GAMMA RAYS—Electromagnetic radiation (light waves, radio waves, and x-rays) of extremely short wavelength. Like all electromagnetic radiation, gamma radiation is carried by particles called photons. Gamma ray photons, however, have very high energies, and consequently are extremely hazardous to health because of the molecular damage that the photons cause as they rip through cells. This damage may especially affect the DNA within cells, causing cell death, transformation to a cancerous state, or mutation. The energy of gamma rays is sufficient to propel them through six inches of lead. Gamma rays are produced when certain kinds of atoms undergo radioactive decay.
See RADIATION; ELECTROMAGNETISM.

GENE—A gene is the basic unit of heredity. Occurring in a chromosome, it is a particular segment of the chromosomal DNA that contains the code for the structure of a particular protein, or which regulates the rate at which such "structural genes" are put into action. All the cells of an organism contains all of the genes characteristic of the entire organism, but only a few genes in any given cell are expressed at any one time. During embryonic development, certain genes are switched into action or out of action, so that the cell or its progeny (following cell division) become more and more specialized for the function they will perform.

Some human hereditary traits are determined by the presence of a single pair of genes (one from each parent), and thus tend to exist in an all-or-nothing fashion. These include trivial traits, such as whether an individual's earlobes are attached or pendulous, and vitally important traits, such as whether one's red blood cells distort their shape under certain conditions—a phenomenon common to diseases such as sickle cell anemia and thalassemia. Many human traits are under the control of several pairs of genes and thus may exist in a great variety of intermediate forms. These include height and skin color.

At the molecular level, a gene is a linear sequence of the four bases adenine (A), thymine (T), guanine (G), and cytosine (C)—the four letters of the genetic alphabet, each acting as one rung of the DNA ladder. A particular triplet, or combination of three of these bases, specifies one of the twenty amino acids that must be chained together to make the gene's product, a protein molecule. For example, ATA is the code for the amino acid tyrosine; GAC codes for leucine. The "non-structural" genes contain regulatory information, controlling the rate at which a given gene is expressed or completely blocking its expression.

See GENES, HOW THEY WORK.

GENE FLOW—Immigrants from one population to another are the vehicles of gene flow.

See GENE POOL.

GENE POOL—When evolutionary theorists talk about a species, they sometimes find it convenient to think not about many individual organisms, but about the aggregate of all the genes possessed by the members of the species. This is a gene pool. Often the term applies only to a given population of the species—a group in which all the members have contact and may potentially interbreed.

Evolution, then, is a change in the frequency with which any gene occurs in a gene pool. This can be the result of the gene conferring a survival advantage (individuals having the gene live longer, reproduce more successfully, or both) or a disadvantage. Gene frequency can also change through immigration, when members of one population move in with another, bringing along new genes, and begin interbreeding with it. This phenomenon is called gene flow.

See EVOLUTION THEORY.

GENES, HOW THEY WORK—For some strange reason, many biology books and articles explain the significance of heredity and genes by saying that they determine such things as eye color, skin color, stature, and other trivial qualities. They neglect to add that genes also determine that you have two legs, that your head will be on top of your neck, that your heart will be connected to your blood vessels, and that you can talk, among other things. Besides this, genes govern such obscure events as the absorption of oxygen into the bloodstream, the extraction of energy from food, and the way in which nerve cells store memories. In other words, genes govern everything from the tiniest molecular interaction within a cell to the shape of your whole body.

A detailed knowledge of how genes work, then, is fundamental not only to understanding why human beings are not built like geraniums, but to figuring out ways to repair or prevent hereditary diseases. In the last century, Gregor Mendel determined that hereditary traits are governed by some kind of particle in the body. In the first half of this century, others figured out that these particles, called genes, resided in the DNA of which chromosomes are made. But still, nobody knew exactly what genes did and how. The first major advance in this regard began in 1942 when George Beadle, an American geneticist at the California Institute of Technology, was doing experiments on fruit flies—the guinea pigs of early genetics research. Beadle had two strains of fruit flies from the same species—one with normal and one with abnormal eye color. He found that if he took some cells from a normal larva and transplanted them into an abnormal larva, the abnormal larva would develop the normal eye color.

Breeding experiments had shown that the fruit fly's eye color was a genetically governed trait, but Beadle had no idea what the gene actually did to make the eyes the right color. His experiment showed that something in the transplanted cells (which had normal genes) must have made up for the defective or missing gene in the recipient larva. Since genes had never been seen to migrate from cell to cell, the only thing that could have gotten into the fly's

eye and made up for its defective gene was a chemical
produced in the cells of the normal larva and released to
the defective larva at large. Beadle reasoned that this
must have been the same chemical that the eye cells of the
abnormal larva would have produced on their own if they
had had the normal gene.

A gene, Beadle guessed, causes the cell to make some
specific chemical—some molecule it needs to carry out its
functions—or, in the case of the fruit fly eye-color gene, to
endow the eye cell with a particular color. It was a brilliant insight, and soon afterward Beadle was joined by
Edward Tatum, an American biochemist, who helped develop the idea further. Tatum knew that the chemical reactions in living cells are mediated by enzymes. These are
protein molecules whose special structure causes other
chemicals to react in specific ways. Genes, the two scientists suspected, caused a cell to carry out specific chemical
reactions.

Beadle and Tatum decided to test their idea on an even
simpler organism than the fruit fly—bread mold. Molds,
like all living organisms, grow by drawing nutrients from
their environment and either using them directly or converting them to other substances that are needed. Enzymes do the converting inside the cells of the mold. First
the scientists zapped mold spores with x-rays, a procedure
others had found would randomly damage genes, creating
mutant cells or organisms. Then they tried to grow the
mutant molds they had created on the usual layer of nutrients. When they found a mold that failed to grow, they
theorized that it had lost the gene for an enzyme needed
to convert a nutrient from the layer into some substance
essential for the mold. By adding missing substances one
at a time to the nutrient layer, and seeing whether the
defective mold started growing, the scientists were in
many cases able to figure out which gene was damaged.
For example, Beadle and Tatum found one mold that
didn't grow until they added arginine, a component of protein molecules that is needed by many cells in order for
them to grow. When they mixed a little arginine in the

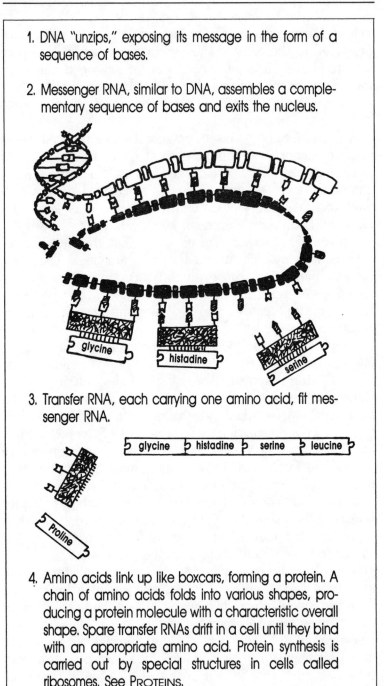

1. DNA "unzips," exposing its message in the form of a sequence of bases.

2. Messenger RNA, similar to DNA, assembles a complementary sequence of bases and exits the nucleus.

glycine

histadine

serine

3. Transfer RNA, each carrying one amino acid, fit messenger RNA.

glycine histadine serine leucine

Proline

4. Amino acids link up like boxcars, forming a protein. A chain of amino acids folds into various shapes, producing a protein molecule with a characteristic overall shape. Spare transfer RNAs drift in a cell until they bind with an appropriate amino acid. Protein synthesis is carried out by special structures in cells called ribosomes. See PROTEINS.

nutrient layer, the mold grew nicely. Apparently, the mold had simply absorbed the arginine from the nutrient layer. It didn't have to make arginine from other nutrient molecules.

Through many such experiments, Beadle and Tatum proved by 1944 that genes code for enzymes, and that each gene codes for one specific kind of enzyme. Today we know that some genes do not code for the manufacture of any substance, but instead act as regulators of genes that do code for enzymes.

Once it became clear that genes carry the blueprints for making protein molecules, the next step was to figure out exactly how they do this; in other words, to figure out what the molecular structure of a gene was. The atoms that make up DNA, it was clear, had to be arranged in a way that carried some kind of information.

Early work on the chemical nature of DNA had established that it consisted of alternating sugar and phosphate units, forming a kind of backbone from which four other kinds of subunits projected like ribs. These four smaller molecules are called adenine, thymine, cytosine, and guanine, which are usually abbreviated to their initial letters. Meanwhile, studies of protein structure had shown that all protein molecules were also made of a long chain of subunits, called amino acids, of which there were twenty different types. The chains, often containing hundreds or thousands of linked amino acids, were usually folded up into characteristic shapes determined by chemical attractions between atoms within the chain. Some protein chains fold themselves into a ball. Some fold up to take the shape of accordion pleats. Most assume a combination of ball, pleat, sheet, and other shapes. It is the shape of the fully folded-up protein, and the chemical nature of the amino acids on the outside of this structure, that determine the protein's function.

DNA, then, is a chain of bases whose sequence determines the sequence of amino acids in a second chain, a protein. It quickly became apparent—though it was not quickly proven—that the genetic code was linear and that the sequence of bases in DNA (A-T-G-C-C-G-A-T, for example) was somehow translated into the sequence of amino acids in a given protein. Through the 1950s many laboratories around the world were working on DNA and related problems. James Watson and Francis Crick at England's Cavendish Laboratory figured out the double helix structure that allows DNA to replicate itself. Other biologists were studying RNA, a related kind of nucleic acid that seemed most common not in the cell nucleus, where DNA always stayed, but near tiny structures within cells called ribosomes. It was learned that ribosomes were protein-assembly plants. Then came an observation by Alfred Hershey, an American molecular biologist, that when viruses attack bacterial cells and dump their DNA into these cells, a small amount of RNA is suddenly synthesized in the bacteria. Another research group noticed the same phenomenon, but extracted the newly made RNA and analyzed it. It turned out to very closely resemble the DNA of the infecting virus.

One Sunday afternoon at a scientific meeting in Cambridge, England, the pieces fell together. Crick was discussing the various findings relating to DNA and RNA with Sidney Brenner, a British molecular biologist, and François Jacob, a Frenchman working in the same field. It dawned on all of them how the system works: DNA keeps the genetic message safe in the nucleus. RNA transcribes the message from the DNA and floats out of the nucleus (which has pores in its wall) to meet up with a ribosome, the protein factory. Somehow the ribosome reads the code in the RNA and assembles the final product, a chain of amino acids that, when finished, folds up into a proper protein. The concept was boiled down to what Crick called the central dogma of genetics: DNA makes RNA makes protein.

The only element missing from that picture, which turned out to be correct, was an understanding of how the four-base alphabet of DNA got translated into the twenty-amino-acid alphabet found in proteins. That would come in the early 1960s. Crick and Brenner and their associates, through a series of brilliant experiments, showed that the code for any given amino acid consisted of a set of three adjacent bases. The DNA code for the amino acid lysine, for example, is T-T-C. If the next three bases are A-T-G, the amino acid being specified is tyrosine. Brenner and Crick realized that for a code like this to work, there must be a clear starting point (so that the triplets of bases are correctly grouped), and that the gene can be read in only one direction thus, A-T-G is tyrosine but G-T-A is histidine).

From 1961 through 1966, Marshal Nirenberg of the United States National Institutes of Health and Severo Ochoa of New York University raced one another to decipher the codes for each of the twenty amino acids used to make proteins. Their methods were laborious, but eventually the complete code had been cracked. The four bases of DNA can be arranged in sixty-four different groups of three, or triplets—more than enough to code for twenty amino acids. As it turned out, many amino acids were coded for by two, three, four, and even six different triplets. Of the sixty-four possible combinations of DNA bases, all except three code for only one specific amino acid. These three are genetic "nonsense" sequences and code for nothing.

After the genetic code had been cracked, one of biology's most dramatic revelations became obvious. Every living organism on Earth—be it a human being, a trout, an oak, a bacterium, or anything else—employs the same genetic code. All use the same four bases grouped in the same triplets to specify the same twenty amino acids. It is perhaps the deepest confirmation of the theory of evolution that could be imagined. And it is the reason why genetic engineers today can talk so easily about transferring genes from alfalfa to corn and from people to bacteria.

See MENDEL; DOUBLE HELIX; DNA, HOW ITS ROLE WAS DISCOVERED; WATSON; CRICK.

GENE SPLICING—The insertion or deletion of genes from an organism's naturally occurring set of genes.

See RECOMBINANT DNA; GENES, HOW THEY WORK.

GENE THERAPY—This is a potentially new form of treatment for human diseases caused by the inheritance of a defective gene. It is an effort to implant in the patient's cells "good" versions of the gene that is absent or defective. The implanted genes are laboratory-made copies of normal human genes. The method remains experimental, but has the potential for helping millions of people suffering from hundreds of diseases that result from single-gene defects, such as multiple sclerosis and muscular dystrophy.

Since genes and cells are too small for the use of surgical methods of implantation, gene therapy makes use of a natural ability of certain viruses to ferry genes into cells. Certain viruses consist of a DNA molecule that carries a few genes packaged in a protein coat. By itself a virus can do little, for it lacks the machinery for putting its genetic code into action. It must invade an ordinary cell and borrow the cell's machinery even to reproduce. Most viruses kill the cell in the process of doing this. One group of viruses, however, does not, and these are the ones that molecular biologists have chosen to implant therapeutic genes.

These special viruses carry their genes in the form of RNA, not DNA. Instead of just dumping their genes into the nucleus of the cell they invade, however, these viruses have an enzyme that causes the RNA to be converted to DNA. They also have enzymes that cut open the DNA of the invaded cell (also called the "host" cell) and splice the viral DNA into it. The cell then automatically makes new viruses, and instead of being killed, passes the viral genes on to its descendants just as if they were normal cellular genes.

Molecular biologists have found ways to remove most of the genes of such viruses (including those needed to make new viruses), and to splice in human genes in their place. When the virus infects a cell, it then inserts the human gene, and in effect, self destructs. From this point on the new gene acts as if it were a normal inhabitant of the cell. When the cell divides, it is duplicated along with the other genes, and one copy is passed into each of the progeny cells.

Although gene therapists could infect patients with the gene-carrying virus and let the new gene take up residence wherever it may, they are reluctant to try this yet. The viruses might infect cells in which the new gene's message might interfere with specialized cell functions. Instead, the doctors want to infect only selected tissue, such as the bone marrow. There is already an established procedure for removing marrow cells and reimplanting them. In this procedure, doctors poke a needle into the bone and suck the cells into a big hypodermic syringe. The marrow cells can then be exposed to the virus and injected back into the patient. It is known that marrow cells will take up residence in the bones and continue dividing for many years. And since marrow cells give rise to blood cells, the patient with a blood disease caused by a defective or missing gene, such as sickle-cell anemia, may eventually have a fair number of circulating blood cells carrying the good gene. The doctors studying gene therapy do not think it is necessary to put a new gene into every cell in the body. Many genetic diseases result from the absence of a chemical, such as an enzyme, that would have been made by the defective gene, but in many cases it turns out that small doses of the missing chemical are enough to prevent the development of the disease. A few circulating blood cells, for example, may be enough to produce the missing substance and circulate it through the body.

See RECOMBINANT DNA; RETROVIRUS; GENES, HOW THEY WORK.

GENETIC CODE—Genes work by carrying information encoded in the form of various small molecules bound in a fixed linear sequence, like pearls on a string. Other molecules can recognize this sequence and cause protein molecules to be assembled according to it.

See GENES, HOW THEY WORK.

GENETIC DRIFT—This is one of the ways in which evolution can happen, along with natural selection and some other mechanisms. Some theorists think genetic drift may play an even bigger role in evolution than natural selection. Whereas natural selection favors traits that improve an individual's adaptation to its habitat, genetic drift favors nothing except change. It can cause a completely nonadaptive trait to evolve, and may be part of the reason why human beings evolved into different racial groups.

In every population of a species there is always a large amount of adaptively neutral genetic variability. Examples of this among human beings include eye color, fingerprint type, and facial features. As long as individuals have a wide choice of potential mates, they are unlikely to choose those with one particular trait over any others. Thus, no particular trait becomes the standard for the species as a whole.

In small, isolated populations, however, the choice is much more limited. A trait that was rare in the large population may, for purely chance reasons, be common in the small group. And even if the small group begins with diverse traits, some chance event—such as an accident or epidemic—may wipe out much of the variety. For example, people with the skinfolds around the eyes (known as epicanthic folds, common to most Asians), may have been a rarity in the large, main population of early human beings, but may have been peculiar to one family that happened to break away. Or the epicanthic fold may have become common only after some random event wiped out most persons with other eye features. As the surviving population grew in isolation, the new trait would then have become standard simply because there was no alternative gene in the population. Evolution caused by these random events is called genetic drift.

See EVOLUTION; NATURAL SELECTION; DARWIN.

GENETIC MATERIAL—This is simply a catchall term for the chromosomes or the DNA molecules of which chro-

mosomes are made. It is often used when the length of
DNA in question is less than that of a whole gene—an
entity called a gene fragment.

GENOME—The full complement of genes belonging in
each cell of any given species. The human genome, for example, includes not only the genes for eye color and
height, but those for two-leggedness, a four-chambered
heart, lungs, and many others.

GENOTYPE—The specific set of genes in each cell of an
individual. The term is used in contrast to "phenotype,"
which refers only to the traits actually manifested in the
developed organism. A person's genotype may include
genes for traits that are not manifested either because
other genes dominate or because environmental conditions
prevent their manifestation. In years past, many Japanese
were phenotypically short in stature, but are now believed
to have had a tall genotype. Poor nutrition and frequent
disease were the environmental effects that prevented full
expression of their genetic potential.

See PHENOTYPE.

GENUS—A group of one or more species that is deemed
to be closely related in evolutionary terms. Ordinarily,
several species make up a genus and several genera (the
plural) make up a family, a still wider classification. The
term genus can be confusing, since biologists normally use
it as part of the species name. For example, human beings
belong to the genus *Homo* and to the species *Homo sapiens*. Quite often the species name (the *sapiens* in this
example) is not unique, and may be found in many genera.
Someone who found a new species of *Australopithecus*,
for example, might choose to call it *Australopithecus sapiens*. This would not imply a close relationship with
Homo sapiens, since one must always consider the species
name as a subset of the genus name.

See SPECIES; LINNAEUS (who set up this system);
TAXONOMY.

Era	Period	Epoch	How Long Ago It Began. (Myr = Millions of Years Ago)	Typical Events
C E N O Z O I C	Quaternary	Recent	10,000	Rise of civilization
		Pleisto-cene	2.5 Myr	First *Homo*
	Tertiary	Pliocene	7 Myr	Origin of hominids and other modern mammal types
		Miocene	26 Myr	Flowering plants begin dominating land with mammals, birds
		Oligocene	38 Myr	
		Miocene	54 Myr	
		Paleocene	65 Myr	
M E S O Z O I C	Cretaceous		136 Myr	Last wave of dinosaurs; origin of flowering plants
	Jurassic		190 Myr	Cycads and conifers dominate land as dinosaurs are abundant; origin of birds
	Triassic		225 Myr	Origin of dinosaurs; primitive mammals
P A L	Permian		280 Myr	Expansion of reptiles; amphibians decline; trilobites die out

E	Carbonif-erous			
O		Pennsyl-vanian	325 Myr	Coal-forming forests of
Z		Mississip-pian	345 Myr	ferns and other primi-tive plants; amphibians dominate; reptiles be-gin
O				
I				
C				
	Devonian		395 Myr	Bony fishes in seas; origin of amphibi-ans; first seed plants
	Silurian		430 Myr	Arthropods and primi-tive plants invade land
	Ordovician		500 Myr	Origin of ver-tebrates (jawless fishes)
	Cambrian		570 Myr	Marine in-verte-brates, al-gae in seas
	PRECAMBRIAN		???	Origin of life

GEOLOGIC AGES—Geologists, paleontologists, and others have found it convenient to divide Earth's past into segments, most of which stretch over several million years. These segments have usually been defined on the basis of characteristic groups of fossils found in sedimentary rocks that formed during that time. For example, the Cretaceous period, which ended 65 million years ago, is defined by having certain dinosaur fossils in it. The end of the Cretaceous period is defined as the time when the dinosaurs became extinct. Younger rock strata (usually lying above older strata) that no longer contain dinosaur remains are considered to be in the next period, the Tertiary. In practice, the definition of the various intervals is more complicated than this suggests, but the principle is the same.

Accompanying this is a table of the major geological intervals recognized by most scientists today. The dates are often estimates, some of which have changed in recent years and some of which could change again.

GERM—This word has several different meanings, but all are related to the root meaning, which refers to a thing that is capable of growing into something larger. The germ in a seed, for example (such as wheat germ), is an embryo of the new plant. The germs that cause disease are capable of multiplying into millions more as they afflict the body. In this meaning, "germ" is not really a scientific term; researchers prefer to be more specific, classifying a germ as a bacterium, virus, protozoan, or some other type of microbe. Often, when scientists don't know the type of microbe causing a disease, they will still prefer to speak of it as an "infectious agent" rather than a "germ." A third use of "germ" is in referring to the body's reproductive organs in the ovaries or testes. These contain cells that divide to produce eggs (or ova) and sperm. The cells that do this are sometimes called germ cells.

GERM THEORY—Perhaps it isn't so surprising that it took science more than two thousand years to figure out that germs cause diseases. After all, many people today still can't quite get over the idea that colds are caused by getting soaked in the rain or by not bundling up in winter. Viruses are, of course, the culprits. But since most people have never seen a virus and many people do get colds in the winter, the germ theory often just doesn't seem persuasive.

Not until barely a century ago did scientists establish the germ theory. It had to wait for Robert Koch to do the experiments in 1876 that first linked a specific microbe, a bacterium, to a specific disease, anthrax.

For most of those two thousand years until Koch's discovery, thinking people accepted the so-called miasmic theory of disease causation. This had its origin around the fifth century B.C. in an idea Hippocrates offered to explain epidemics of disease. It had always been perfectly obvious that large groups of people could suddenly come down with the same disease. And intelligent people naturally looked for some cause that could touch many people simultaneously. Hippocrates attributed epidemics to an unfortunate conjunction of weather conditions. Somehow, it seemed, air could turn bad, and, of course, many people could breathe the same air.

In the Middle Ages, the idea was modified slightly to suggest that noxious vapors from swamps or refuse heaps could turn the air bad. In southern Italy one disease even got the name "malaria," which means bad air. The early association of malaria with swamps was a good one; centuries later it would be shown that mosquitoes from swamps carried the microbe that causes malaria.

An even better idea emerged in the sixteenth century when Girolamo Fracastoro, an Italian physician, speculated that epidemic diseases were spread by tiny particles—he called them seminaria, or seeds—that could be passed directly from one person to another through contaminated objects or through the air. Fracastoro even guessed that these seeds were self propagating. It was a brilliant insight, but without hard evidence it remained only a speculation.

Only in the 1800s did several observations and experiments finally lead to the modern germ theory. John Snow, an English physician, made a classic epidemiological investigation during an 1854 cholera outbreak in London. He mapped the homes of stricken people and found a surprising cluster around one particular well where many people went to get water. Snow ordered the well closed and the epidemic subsided. The cause had to have been something in the water, but still nobody knew what it could have been.

In 1876 Koch, a German bacteriologist, collected matter from the sores of cattle with anthrax, a disease of hoofed animals. He smeared this material over a layer of a special nutrient that he had devised, and observed colonies of tiny organisms growing on the nutrient. He proved that the organisms, bacteria, were the cause of the disease by giving some to healthy animals and observing that they developed anthrax. Koch's methods remain the standard way of proving whether microbial organisms cause a given disease, and what kind of organisms they are. With these methods, Koch subsequently identified the cause of tuberculosis, the greatest scourge of the time.

Koch also worked on malaria, but the disease did not yield its secret until Ronald Ross, a British doctor in the Indian Medical Service, discovered that the malaria parasite (not a bacterium) inhabited a certain species of mosquito and was transmitted when the insect bit people. Malaria, the quintessential miasmatic disease, turned out not to be contagious after all. But the germ theory had begun to be broadened to include the transmission of disease-causing organisms by a variety of means in addition to direct contact.

In the twentieth century, the germ theory has been further broadened to include not only bacteria and microscopic parasites but viruses as well. Although not living organisms in the usual sense, viruses are transmitted in much the same way as other "germs."

See KOCH; VIRUSES; BACTERIA.

GESNER, CONRAD, 1516–1565, Swiss zoologist— Conrad Gesner, a minor major figure in science, was nonetheless the most important zoologist to appear in the two thousand years between Aristotle in the third century B.C. and Linnaeus in the eighteenth century A.D. Although he was a physician practicing in his native Zurich, Gesner compiled the monumental *Historia Animalium,* or History of Animals, a five-volume work describing every species of animal known to the Renaissance world. Reportedly, Shakespeare had an English translation and relied on it for information about animals for his plays. Gesner also wrote, but never published, a *Historia Plantarum*, which would have done for botany what his other book did for zoology.

Although Gesner was not terribly selective—he tossed mythological tales into his book along with direct observation—he was, in a sense, the first science writer. He published less than a century after Gutenberg had invented movable type and even included many illustrations in his books so as to make the information more accessible to nonscientists.

Gesner also published the first well-illustrated book on fossils, and compiled a catalog of every book published in Greek, Latin, and Hebrew since the invention of printing. For this last work, he is claimed as the founder of bibliography.

GLASHOW, SHELDON, 1932– , American physicist. See GRAND UNIFIED THEORIES.

GONDWANA—This is the name of one of the two continents that existed on Earth about 380 million years ago. The other was called Laurasia. By about 250 million years ago the two had joined to form a single supercontinent called Pangaea. Pangaea then broke up, between 200 million and 135 million years ago, separating the two continents once again. Gondwana consisted of the land that is now South America, Africa, Antarctica, Australia, and India. It broke up gradually, and its pieces drifted to their present positions on the globe. Laurasia included North America and Eurasia.

See PLATE TECTONICS; WEGENER (who first developed the idea).

GOULD, STEPHEN JAY, 1941– , American evolutionary biologist. See PUNCTUATED EQUILIBRIA; MODERN SYNTHESIS.

GRADUALISM—A slow and steady mode of evolution resembling that espoused by Darwin, but now held in some doubt.

See PUNCTUATED EQUILIBRIA; EVOLUTION THEORY; DARWIN.

GRAND UNIFIED THEORIES—A general term, often abbreviated GUTs, for theories in physics that try to show that today's forms of matter and energy are descended from a single primordial entity that existed at the very start of the Big Bang. Physicists suspect that in the first billionth of a billionth of a second after the start of the Bang, a single phenomenon (some call it a superforce) split into different manifestations, yielding the particles that make up atoms (quarks, electrons, and such) and the four fundamental forces that operate today—gravity, electromagnetism, and the two forces inside atomic nuclei, called the strong force and the weak force. These four forces account for all forms of energy.

Although scientists since Einstein have labored to find the mathematical relationships that would link the entities that became separated in the Big Bang, no one has achieved their total unification. There has, however, been progress. In recent years three physicists, Sheldon Glashow, Abdus Salam, and Steven Weinberg, have shown that electromagnetism (the force that operates in electricity) shares a common ancestry with the weak nuclear force, which operates when atomic nuclei decay. Physicists conjecture that this unified force, called the electroweak force, shared a common ancestry with the strong nuclear force. Various rationales are offered to support this, but physicists do not agree on any one of them. Linking the electroweak force with gravity— by far the least understood of forces—remains a distant goal.

See BIG BANG; FORCE; ELECTROMAGNETISM; GRAVITY; STRONG FORCE; WEAK FORCE.

GRAVITY OR GRAVITATION—A force that operates between any two objects having mass, causing them to attract one another. Gravity is the force that holds planets and stars in their spherical shapes and keeps solar systems and galaxies organized. It is one of the four fundamental forces in nature, the others being electromagnetism and the two nuclear forces called the weak and the strong force. Nobody really understands exactly how gravity works, but there are two rather different interpretations that physicists use.

One is that gravity, like the other forces in nature, results when the attracting bodies exchange particles carrying this force. The gravity particle is called a graviton. Nobody has ever found a graviton, but several scientists are hoping to detect gravity waves—sudden changes in the strength of gravity coming from some point in the universe where, perhaps, a star is collapsing into a black hole.

The second interpretation of gravity, offered in Einstein's relativity theory, is that it is not really a force in its own right, but a kind of illusion resulting from the fact that space is curved in the vicinity of any object with mass. The more massive the object, the steeper the curvature. A two-dimensional analogy pictures a heavy ball (the object with mass) resting on a stretched rubber sheet (the space around it). Just as the ball warps the otherwise flat rubber sheet, so mass warps space. A lighter ball, rolling near the first, would be caught in the curvature and roll toward the heavier ball. This is only a two-dimensional analogy, however, and the concept of curved space, though it makes sense in two dimensions, is harder to envision in three dimensions. Nonetheless, the point of the analogy is that the heavier ball does not act directly on the lighter ball, but instead acts on the space around it. The lighter ball simply follows the resulting curvature of the space.

Reconciling these two views of gravity is one of the goals of modern physics.

See FORCE; NEWTON (who first recognized the universality of gravity); RELATIVITY; EINSTEIN; BLACK HOLE.

GREENHOUSE EFFECT—The Earth's atmosphere plays much the same role for the Earth's surface as does

the glass in a greenhouse. It lets sunlight in and prevents
some of the resultant heat from radiating back out. One of
the gases in air that helps produce this effect is carbon
dioxide (CO_2), and there is fear that if its carbon dioxide
concentration increases, the atmosphere will become
even more effective in trapping heat. A rise in the
Earth's average surface temperature of only two or three
degrees, caused by such trapped heat, would shift the
boundaries of some climatic zones toward the poles, and
melt enough polar ice to raise the level of the oceans by
several feet. Most of the world's coastal cities would be
destroyed.

As it happens, burning fossil fuels produce carbon dioxide, and the amount of this gas in the atmosphere does
appear to be increasing, as measured high above the Pacific Ocean. It is not yet clear whether any climatic effects
have come from this, although there is evidence that the
sea level is indeed rising slowly. Much of the greenhouse
effect may be counteracted by the increasing amount of
particulate pollution in the upper atmosphere. The particles reflect some sunlight back into space before it can
reach the ground, and some experts think that this phenomenon may be offsetting the greenhouse effect.

There is no practical way to prevent the production of
carbon dioxide as long as the world relies as heavily as it
does on coal and oil. Even the burning of wood produces
carbon dioxide. In fact, the better most pollution control
systems are, the more completely they convert toxic hy-
drocarbons into water vapor and carbon dioxide. And the
better the pollution control, the smaller also is the amount
of particulates put into the atmosphere.

One thing that may retard the accumulation of carbon
dioxide is the plant life of the world. Plants need carbon
dioxide. They take it in and combine it with water to make
sugar, giving off oxygen as a waste product. There is evidence that increasing carbon dioxide levels are causing
some forests to grow faster. If the effect is worldwide—
especially in the oceans, where algae, the Earth's main
oxygen producing plants, live—plants may come to our

rescue. It may just be that the more carbon dioxide we put into the atmosphere, the faster they will take it out. Or it may not be. The billion or so people who live near seacoasts may still have to move inland.

See PHOTOSYNTHESIS.

GRIFFITH, FREDERICK, 1879–1941, British bacteriologist. See DNA, HOW ITS ROLE WAS DISCOVERED.

HALF-LIFE—This is a term originally developed as a measure of the rate at which radioactive elements decay, but which is useful in describing similar statistically random events of any kind. The half-life of the most common form of uranium is 4.5 billion years. This means that over that period of time, half the nuclei of all the uranium atoms in the world will spontaneously decay, with each atom emitting a neutron and a burst of energy. As it happens, 4.5 billion years is also the estimated age of the Earth. This means that when our planet was formed, it had twice as many uranium atoms in it. When another 4.5 billion years have passed, the Earth will have half as many as it now has, or one-quarter the original amount.

Of course, if some uranium atoms happen to find their way into the fuel of a nuclear power plant, they will be subject to a much faster decay rate because their breakdown can be triggered by bombardment with neutrons.

Uranium, like all other unstable elements or particles, follows the half-life rule because under natural conditions on the Earth, the phenomena that break up an unstable particle are essentially random. The natural event that breaks up a uranium nucleus, whatever it may be, has only a 50 percent chance of happening in 4.5 billion years. Thus, only half the atoms of uranium are likely to experience it. The atoms that escaped during the first 4.5 billion years of the Earth's history are, in effect, throwing the dice once again in the next 4.5 billion years, and again, there is only a one-in-two chance that the decay-causing event will occur in each of them.

Some natural atoms have much shorter half-lives. Radium, for example, has a half-life of only 1,622 years. Some of the particles created in particle accelerators (atom smashers) have half-lives of only a few millionths or billionths of a second.

See RADIOACTIVITY; DECAY (RADIOACTIVE); ATOMIC PARTICLES.

HALLEY, EDMUND, c. 1656–1743, English astronomer. See COMET; NEWTON.

HARVEY, WILLIAM, 1578–1657, English physician— William Harvey was the first to discover that blood circulated in the body—that the heart pumped it out through the arteries and received it back through the veins. Until this discovery, most people clung to the ancient teaching of Galen, fourteen hundred years earlier, that blood originated in the liver and spread through veins to all parts of the body, or "sloshed about" in a kind of tidal ebb and flow. In the old view, the arteries contained blood and air from the lungs, although Galen did insist that there was only blood. The movement of blood was thought to result from a pulsing in the blood vessels.

Harvey's discovery, published in 1628, straightened out the confused explanation of how arteries, veins, blood, and the heart were related. Through a series of experimental demonstrations, Harvey showed that there was one coordinated circulatory system with the heart at the center as pump. This was a twofold advance. First, it signaled the beginning of the end of hidebound faith in Galen's unscientific teachings, and launched the modern era of scientific medicine. Second, it revealed an important physiological mechanism linking many other systems of the body. For example, once it was clear that blood circulated through the lungs, a means was established for understanding how oxygen was carried throughout the body. Harvey also drew in the digestive system, showing how circulating blood carried nutrients to all other tissues.

There was, however, one embarrassing gap in Harvey's story of circulation. He had the blood flowing out through the arteries and back through the veins, but he could not produce any evidence of how the blood got from the end of an artery into a vein for the return trip. All the other points of Harvey's new theory were backed up by experimental data, gathered mostly through dissections of animals, some living long enough for Harvey to examine their beating hearts. Lacking anything better than a magnifying glass, Harvey could not see the microscopic blood vessels called capillaries that link arteries to veins. He calculated, however, that if there were no connection, the heart would quickly drain the veins of all blood and burst the arteries. Since this plainly doesn't happen, he asserted that even though they couldn't be seen, there had to be tiny blood vessels that linked arteries with veins.

Harvey's life was one of affluence and privilege. His father was a prosperous businessman. After completing medical school Harvey married the daughter of King James's personal physician, a connection that would eventually pay off. But not at first. When Harvey tried to get an appointment as physician to the Tower of London, where he could meet and treat the many distinguished men imprisoned there, he failed. He then developed a private practice and catered to many of the most eminent citizens of his day. Sir Francis Bacon, for example, was one of his patients. Finally, at the age of forty, he was appointed physician extraordinary to King James I. When James died despite Harvey's care, the new king, Charles I, continued Harvey's appointment. King and doctor became good friends. The king helped Harvey in his animal research by allowing him to experiment on deer in the royal game reserves.

Although Harvey suffered painfully from gout and kidney stones, he lived until the age of eighty, when he died suddenly of what was probably a stroke—a blood clot in the circulatory system supplying the brain.

HEAT—A form of energy that, on a molecular level, results from the motion of the molecules as they rotate and vibrate within a circumscribed region. When an object is heated, its molecules vibrate faster, causing the object to expand slightly. As the object cools, the molecules move less and the object shrinks.

Heat can be transmitted from regions of higher temperature to regions of lower temperature in three ways: conduction, convection, and radiation. In conduction, the molecules of the hotter substance literally bang against the molecules of the colder region, setting them in motion. Since energy is transferred, the hotter region cools, because its molecules have less energy with which to vibrate. In convection, a heated fluid expands, becomes less dense than the cooler fluid around it, and rises. Radiation can transfer heat energy over indefinitely long distances and through space. As the electrons in a heated atom absorb energy or, as physicists say, as the atoms become excited, they jump to higher shells as they orbit the atomic nucleus. But since the higher shells are unstable positions, the electrons quickly drop back to a lower shell, emitting a burst of electromagnetic energy in the process. If the rate of jumping between shells is low, the electromagnetic radiation carrying this energy is in the infrared range of the spectrum, producing heat. As the emitted energy (temperature) increases, the emitted radiation changes frequency and may enter the visible range of the electromagnetic spectrum, producing light we can see. Light bulbs glow because electricity flowing through the filament excites atoms to such rapidly repeating electron jumps that the emitted radiation attains the frequency that our eyes detect as light.

See THERMODYNAMICS (the science of heat); ATOMIC STRUCTURE; ELECTROMAGNETISM.

HEGEL, GEORG W. F., 1770–1831, German philosopher. See HELMHOLTZ.

HEISENBERG, WERNER KARL, 1901–1976, German physicist—One of the founders of quantum mechanics—the theory underlying the twentieth century's deepest insights into the nature of the atom—Heisenberg is best known for formulating his so-called uncertainty principle, which is also called his indeterminancy principle. The principle, explained in more detail below, says basically that there are certain kinds of information about the subatomic world that cannot be learned with great specificity; only probability estimates can be obtained for them. Heisenberg was the first to grasp this profound—and to some people, dismaying—consequence of the fundamental nature of matter and energy, especially the strange dual nature of each as both particle and wave.

Heisenberg, who won the 1932 Nobel Prize for helping to found quantum mechanics, was born at Würzburg, Germany, where his father was a professor. He studied theoretical physics in Germany and then worked in the 1920s under Neils Bohr at Bohr's famed institute in Copenhagen, which was then the world center of research in fundamental physics. Heisenberg then returned to Germany and became director of a succession of theoretical physics research institutes.

See HEISENBERG'S UNCERTAINTY PRINCIPLE; QUANTUM MECHANICS; ATOMIC STRUCTURE.

HEISENBERG'S UNCERTAINTY OR INDETERMINANCY PRINCIPLE—This is probably one of the most widely misunderstood concepts in science. It does *not* mean that you can never be sure about anything. It simply refers to the fact that when you are studying the subatomic particles that make up atoms, you cannot simultaneously measure both the position and the momentum of any one such particle. The reason for this is that the particle is disturbed unpredictably by what you must do to it to measure either quantity. In other words, if you measure its momentum, its position is altered in some unpredictable way, and vice versa. The best you can do is make a statistical statement of the probability that the particle is in a given place when you measure its momentum, or has a given momentum when you determine its location. The reason that subatomic particles behave in this way has to do with the fact that although they are

usually thought of as particles—or little balls of matter—
they also behave like waves, which are not objects themselves but disturbances in a medium. The uncertainty
principle was developed in 1927 by Werner Heisenberg, a
German physicist.

Heisenberg's principle erects what seems to be an im-
penetrable barrier to science's direct probings of the ultimate nature of reality. Since we can only perceive reality
through the medium of our senses, there is a limit to what
we can know. At some point, observation itself can blur
the view. We are left only with mathematics as our tool
for probing further. And those of us who cannot follow the
formulas of mathematics must take on faith much of what
is discovered. Even those who can do the mathematics
must proceed on the faith that reality itself obeys mathematical laws. That faith, however, has been vindicated
many times, not least when the first atomic bomb exploded in the American desert, converting some matter
into energy, just as the formula $E = mc^2$ had said it would
decades earlier. In fact, the seemingly uncritical drive of
many scientists to make an atom bomb can be seen in part
as the desire for a sign from the god of mathematics that
their faith was warranted.

Einstein himself was never very happy about the uncertainty principle. "I cannot believe," he once said, "that
God plays dice with the world." Einstein clung to the hope
that someday scientists would find a way to break the uncertainty barrier, revealing its adoption to be only a tem-
porary expedient in the progress of physics. Half a
century of accelerating progress in physics, however, has
not only failed to break the barrier, but has led practically
everyone in physics to the even stronger conclusion that it
is real.

See ATOMIC STRUCTURE; QUANTUM MECHANICS; REL-
ATIVITY; HEISENBERG; EINSTEIN.

HELIOCENTRIC SOLAR SYSTEM—One of the
great intellectual advances of all time was the rejection of
the ancient belief that the Earth was immobile while the

rest of the universe circled about it. In the sixteenth century, Copernicus proposed that the Earth orbited the sun. In the seventeenth century Galileo used one of the first telescopes to find evidence proving Copernicus right. At about the same time Kepler confirmed Galileo's observation and added more details, showing that the orbit of the Earth around the sun was not a perfect circle but an ellipse.

See COPERNICUS; GALILEO; KEPLER.

HELMHOLTZ, HERMAN VON, 1821–1894, German physiologist and physicist—Hermann von Helmholtz was the last nineteenth-century physicist. In his fervor of rebellion against the philosophy of Kant and Hegel and the notion of vitalism, which imputed an unknowable mystical quality to life, Helmholtz pushed the theories of classical Newtonian mechanics as far as they could go in explaining natural phenomena, including the newly emerging realm of electromagnetism. It is perhaps fitting that he died just before the world of physics underwent the revolution wrought by Albert Einstein's relativity theories and the discovery of the non-Newtonian world inside the atom.

Helmholtz made contributions to several specialties, ranging from physiology—he was trained as a physician—to optics and electromagnetism. He is probably best known as one of several scientists who arrived independently at the law of conservation of energy, which states essentially that energy is neither destroyed nor created, but changed from one form to another. It is a companion law to that of the conservation of matter, which can be traced to the ancient Greeks. Both laws were eventually combined by Einstein, who held that matter and energy can each be changed into the other, but not destroyed utterly.

Hermann Ludwig Ferdinand von Helmholtz was born at Potsdam, near Berlin, and remained virtually housebound for his first seven years because of ill health. His father, a high-school teacher of philosophy and literature, taught young Helmholtz Latin, Greek, Hebrew, French, English,

Italian, and Arabic. He also taught him Hegel's philosophy of deducing conclusions about the natural world from philosophical ideas rather than from observation and experimentation. Hegel's ideas, in turn, derived from Kant, who held that time and space were not real things in the real world, but mental attributes through which people perceive the world. In other words, the order that we perceive in nature is in fact an order imposed by our minds. Helmholtz was to rebel vehemently against this philosophy, insisting, as most scientists do today, that the order they perceive in nature is really there—that nature is in fact supremely and invariably orderly. Science, in this view, is a process of trying to detect and describe that order.

Helmholtz was also to reject the related philosophy of vitalism. This was the view that living things, although made up of matter in various forms, were infused with a so-called vital force, an unknowable property, endowed by God, that coordinated and motivated living organisms. Helmholtz, by contrast, insisted that organisms were nothing more than matter and energy, and that all the wonders of muscular motion, vision, thought, and so forth could be explained by the exquisitely detailed workings of molecules. His was a strictly mechanistic approach.

Young Helmholtz, not wealthy enough to afford a college education, was sent to medical school, which could be had free from the government in exchange for a promise to serve for eight years as an army doctor. Because he showed promise as a scientist, the government released Helmholtz early from his obligation, and he went on to a series of professorial appointments at various German universities.

His early work was on the optics of the eye, which vitalists considered an excellent example of the divine mind at work. Helmholtz showed the eye to be a rather imperfect optical machine. Along the way, he also invented the opthalmoscope, a forerunner of the device eye doctors use today to look inside the eyeball.

As the mechanistic philosophy became stronger in

Helmholtz, he turned increasingly to physics for the ultimate answers to how the world works. He worked on the
phenomenon of electromagnetism discovered by Michael
Faraday and described mathematically by James Clerk
Maxwell, hoping to explain the propagation of electromagnetic waves through mechanical forces that acted in conjunction with "ether," the weightless substance then
thought to pervade all of space. There was even the idea
that electromagnetic waves, including light, moved by
somehow pushing against the ether. Helmholtz's theories
and calculations came very close to explaining the phenomenon of radio waves, which had been discovered by
his pupil, Heinrich Hertz. As it happened, though,
Helmholtz died without completely achieving his goal. In
less than a decade, physics would move away from the
realm of everyday objects and into the very strange world
inside the atom, where Newtonian physics does not apply.
Einstein's ideas would eventually abolish the concept of an
ether, and twentieth-century physics would grapple with
realities far more bizarre.

See CONSERVATION OF ENERGY; EMPIRICISM; EIN-
STEIN; FARADAY; MAXWELL; PART 1, WHAT IS SCI-
ENCE?

HERSHEY, ALFRED, 1908– , American biologist.
See GENES, HOW THEY WORK.

HERTZ, HEINRICH, 1857–1894, German physicist. See
HELMHOLTZ.

HIPPOCRATES, 460 B.C.–377 B.C., Greek physician—
Though he is still widely described as the father of medicine,
almost nothing is known for certain about Hippocrates'
ideas, discoveries, or teachings. Even the facts of his life,
including the dates of his birth and death, are guesses.
Hippocrates is called the father of medicine mainly because
ancient Greek writers called him that. The best clues to his
contributions come from an analysis of the teachings historically associated with the medical school at Cos, where he
taught.

Until the rise of the Cos school and a closely associated school at Cnidus, medicine was largely grounded in magic. Hippocrates' school was instrumental in shifting medicine toward a more scientific course. The followers of Hippocrates claimed that diseases always had specific causes, and that the causes could be learned through a process of rational examination and analysis. This was a big step toward the modern view of disease, but the rational causes implicated by the early Hippocratic physicians had largely to do with the ebb and flow of various "humors" in the body—blood, phlegm, yellow bile, and black bile, according to one view. It would be more than two thousand years before anybody recognized germs. Even the idea of trying to cure a patient was not yet established. The art of Hippocratic medicine consisted mainly of diagnosis and prognosis.

The reluctance to intervene in the course of a disease is reflected in the so-called Hippocratic Oath that some medical schools still administer to their graduates. Among other things it pledges doctors, above all other things, to "do no harm." As it happens, one thing that is fairly certain about Hippocrates is that he is not the source of the oath.

See HARVEY; GALEN; VESALIUS.

HOMINID—At some point in evolution, a species of primate split into two populations. One evolved into the modern apes and the other into modern human beings. All of the individuals on the human side of the division are hominids. Thus the term includes all people living today as well as all of our ancestors back to the common ancestor of the apes and humans. The word hominid is derived from the name for the taxonomic family to which humans belong, the Hominidae. Apes belong to the family Pongidae and are called pongids. Also belonging to the human family are descendants of the original hominids that branched off the main evolutionary line leading to human beings and became extinct.

See HUMAN EVOLUTION; TAXONOMY.

HOMINOID—Although often confused with "hominid," this term has a broader meaning. It refers to all hominids and all pongids, or apes.

See HOMINID; HUMAN EVOLUTION.

HOMO—This is Latin for "human," and not to be confused with the Greek *homo*, which means "same." *Homo* is the genus name for the human species living today and for two extinct species of human being, *Homo habilis* and *Homo erectus*.

See HUMAN EVOLUTION.

HOOKE, ROBERT, 1635–1702, English physicist— There is almost no field of science that does not owe a debt to this man of wide-ranging talents. He not only reached new insights into the processes of nature—even clashing with Newton over who first discovered a certain law of gravitation—but also invented many kinds of instruments for observing and measuring the processes.

Robert Hooke is perhaps best known for his book *Micrographia* (Small Drawings), in which he published his meticulous sketches of what he saw through the instrument he had just invented—the compound microscope. Others, such as Leeuwenhoek, had used a single-lens microscope to make major discoveries, but it was Hooke who conquered the optical problem of grinding and stacking lenses so that one multiplied the magnifying power of the next. Leeuwenhoek was so secretive about how he made his lenses that many people doubted his claims of having discovered a teeming world of tiny animals in a drop of water. Hooke amply confirmed the discovery and probed even deeper, eventually coining the term "cell" for the chambers he saw in a thin slice of cork. He used the term because the chambers reminded him of the monks' cells in a monastery. Nonetheless, Hooke thought cells were simply voids. It was not until the nineteenth century that the modern concept of cells as units of life emerged.

From the very small, Hooke's interests ranged to the

very large. He built the first reflecting telescope, the kind of telescope that is today the most powerful.

Hooke's work with lenses, mirrors, and the behavior of light led to his suggestion that light has wavelike properties. This produced his first clash with Newton, who felt that he had made that discovery first. (Light is now known to have properties both of waves and of particles.) Hooke's other tangle with Newton involved the theory of universal gravitation, the idea that every celestial body exerts a gravitational pull on all the bodies around it. Hooke had written to Newton with his idea, and had also revealed his intuitive guess that the strength of the gravity decreases in proportion to the square of the distance between two bodies. When Newton later published these ideas as his own, Hooke charged plagiarism. Newton replied that whereas Hooke may have been right in his intuition, Newton had established the fact more decisively through his detailed mathematical derivations.

Hooke's other contributions include such inventions as a practical air pump, the iris diaphragm (which is still used on cameras today), the wheel barometer (which gives readings on a dial), and the crosshair sight for telescopes. He also invented a weather clock that simultaneously recorded barometric pressure, temperature, rainfall, humidity, and wind velocity, all on a rotating drum, and suggested that the zero point on a thermometer should be the freezing/melting point of water (as in today's Celsius scale). In the field of mechanics, Hooke invented the universal joint, a device used today to connect an automobile's spinning driveshaft to the axle without breaking when the car bounces up and down.

Hooke even made contributions to architecture and city planning. After the Great Fire of London in 1666, he was Christopher Wren's surveyor, and reportedly helped design some of the new buildings that were put up.

See LEEUWENHOEK; NEWTON.

HORMONE—This term applies to substances manufactured in one part of the body that then travel to another part of the body to produce a stimulating or repressing effect. There are scores of different hormones produced in the bodies of humans and other animals. Some are produced in glands (the adrenal, pituitary, and thyroid glands, testes, and others) that can pour out large quantities of a hormone to circulate in the bloodstream to all parts of the body. Other hormones are produced in individual cells in microscopic quantities, and affect only one or a few nearby cells. Brain cells, for example, communicate with one another by releasing neurohormones that travel from one cell to the next.

Hormones are the messengers of one of the body's main internal communications networks. They play a major role in coordinating events throughout the body, including physical growth, sexual maturation, and, in women, the menstrual cycle. Thus, for instance, the pituitary gland, situated at the base of the brain, produces several important hormones, including growth hormone and the hormones that govern ovulation and preparation of the uterus for pregnancy.

Certain substances in plants are also called hormones, and they too seem to have regulatory and communication roles.

HUBBLE, EDWIN P., 1889–1953, American astronomer—If one person is to be singled out as the originator of our modern concept of the universe, it must be Hubble. He was the first to realize that our own galaxy was not unique in the universe, that there were many others. In fact, he was the first to realize that the universe is almost uniformly strewn with galaxies. And he was the first to realize that all of the galaxies are flying away from one another, that the universe is expanding. All of these realizations emerged from Hubble's painstaking analyses of photographs and other data collected through the giant telescope at Mount Wilson in California. The combined picture that emerged was one of a vastly larger universe than anyone had imagined before, and of a universe that was expanding, or exploding, as it were, from a common point of origin.

Hubble was born in the small town of Marshfield, Missouri. He went to college at the University of Chicago and at Oxford University in England. After serving with the U.S. Army in France during World War I, Hubble got a job at the Mount Wilson observatory in 1919. He worked there until his death thirty-four years later.

Hubble's first major discovery was that a heretofore fuzzy patch of light called the Andromeda nebula was, in fact, a huge, wheeling system of billions of stars rather like our own galaxy, which is called the Milky Way. This was the first hint that other parts of the universe might be like our own, or, viewed in another way, the first hint that our part of the universe was nothing special. Copernicus and Galileo had forced the same realization on the world, but with respect to a much smaller part of the universe, our solar system.

In every direction that Hubble pointed his telescope, he found more nebulae, or fuzzy patches. Detailed analyses of these nebulae gradually revealed that most were galaxies themselves. The universe, he discovered, is a vast soup of giant star systems.

Hubble's most significant discovery emerged in 1929 from his analysis of a peculiar distortion in the light reaching Earth from faraway galaxies. Certain features in the light spectrum were shifted toward the red end, or low-frequency end of the spectrum. This so-called red shift, Hubble realized, was the result of something called the Doppler effect. It applies to sound waves just as to light waves, and is what causes the pitch of a sound to be shifted to a lower register if the distance between the source and the receiver is growing rapidly. The red shift in light from the galaxies, Hubble asserted, must occur because the distance between Earth and the galaxies is growing rapidly. In other words, the universe is expanding. Hubble's assertion was controversial for many years, but today it is generally accepted as a consequence of the event that created the universe—the Big Bang.

See GALAXY; RED SHIFT; DOPPLER EFFECT; BIG
BANG.

HUMAN EVOLUTION—In 1860, shortly after Darwin's bold assertion that human beings had descended
from animals, the shocked wife of the Bishop of Worcester
in England is alleged to have said, "Let us hope it is not
true, but if it is, let us pray it does not become generally
known."

The story may be apocryphal but the sentiment is not.
People then were genuinely troubled by the implication
that their ancestors, however distant, were not people at
all but animals that looked like apes.

Today we are not shocked. Like orphans searching for
our biological parents, we want to know where we came
from, how we got here, and how we are truly related to
the rest of the living world. Major portions of that evolutionary transition have been documented. Except for creationists, who insist upon a literal reading of the biblical
Genesis and its assertion of a supernatural origin of everything, no one doubts that our heritage can be traced back
nearly four million years to little creatures that, as adults,
stood only about as tall as today's five-year-old. In fact,
the transition from these little, prehuman creatures with
brains the size of a chimpanzee's to modern human beings
is one of the best-documented origins of a species in the
known fossil record.

The evidence comes not just from fossils but also from
comparisons of human beings with living apes. It seems
inescapable, for example, that most of our chromosomes—
the genetic structures that encode our heredity—are the
same as those of the most ancient humans. This must be
the case because we now know that our chromosomes are
virtually identical to those of chimpanzees and gorillas.
Even the individual genes residing on those chromosomes—which govern our individual traits—are virtually identical with those of these two animals. The
human gene for hemoglobin (the oxygen-carrying molecule
in blood), for example, is 99 percent identical in structure

to the chimpanzee's hemoglobin gene. This is equal to the genetic difference between chimps and gorillas (the two African apes), and smaller than the difference between chimps and the Asian apes, such as gibbons and orangutans. In other words, we are as closely related to chimps as they are to gorillas, and the three of us—humans, chimps, and gorillas (all of which evolved in Africa)—stand as a definite evolutionary group, well separated from other apes and monkeys.

Our ancestors, however, were not apes—at least not apes of the sort we know today. But today's apes and today's humans do share an ancestor. We are both descended from the same parental species, though no one knows for sure what it was. Millions of years ago that species somehow became separated into two isolated populations. Barred, perhaps by some geographic barrier, from interbreeding, one group evolved its own distinctive features and became the modern apes, while the other became human beings.

It is at this fork in the evolutionary road that the anthropologists who study early humans—paleoanthropologists—mark the beginning of the hominid lineage. Hominids are members of the family Hominidae; this includes human beings and all human ancestors that lived after the evolutionary split from the apes, as well as humanlike creatures that came to evolutionary dead-ends. Species on the ape side of the split are called pongids, or members of the family Pongidae.

RAMAPITHECUS—The oldest fossil primate that has been considered a hominid is called *Ramapithecus*. There has been considerable debate about whether it really was a hominid or whether it was instead a pongid. Some experts suggest that it may have been the common ancestor of both lines. Its fossil remains are relatively scarce—mainly jaws and teeth found in Asia, eastern Europe and, in one case, in Africa. *Ramapithecus* lived from about 14 million years ago until about 8 million years ago. One major challenge to this animal's putative role as a hominid is the biochemical evidence of similarity between modern hu-

mans and modern apes. The amount of difference has been
calculated as the amount that would have accumulated if
apes and humans diverged between three and seven million years ago, which is after the most recent known
Ramapithecus lived.

Whether *Ramapithecus* was a hominid or not, its fossils
do attest to one humanlike evolutionary development. The
jaws and teeth are powerfully built and show wear marks
suggesting that some of these creatures relied on a diet of
tough seeds and grasses. Since apes subsist mainly on
softer fruits and shoots of the deep forest, the fossils of
Ramapithecus indicate that it was built for the open
plains—the very environment in which other evidence
suggests humans arose.

AUSTRALOPITHECUS AFARENSIS—This is the
famous Lucy, found in Ethiopia by the American paleoanthropologist Donald Johanson. Actually, Lucy is only
one of many examples of this species, the earliest un-
disputed hominid. The specimens are all from East Africa,
and range in age from just under 4 million years to around
2.5 million years. Because no hominid species is known
from the 4 million years before *Australopithecus afarensis*, nobody knows how this species arose. *Australopithecus afarensis*, however, was clearly well along on one
of the most significant transitions in all of human evolution—that from quadrupedalism (walking on four legs) to
bipedalism (walking on two legs). We know this because
we have fossils of the hip and knee bones of this species,
which were clearly built for two-legged walking, although
it probably was a shambling gait and not fully erect
striding of the sort modern humans exhibit. We also have
a remarkable set of footprints of *Australopithecus afarensis*, preserved in hardened volcanic ash in Tanzania. The
footprints, though small, in keeping with the three-and-a-
half to four-foot height of an adult *A. afarensis*, look quite
modern.

Unlike the case with *Ramapithecus*, we have many
parts of the body of *A. afarensis*. From the neck down
these creatures look much more modern, although

smaller, than one would expect for their antiquity. Their heads, however, were quite primitive, with jaws and teeth much like those of an ape and a brain no bigger than that of a chimpanzee, with a volume of about 400 cubic centimeters. (Modern human brains, by comparison, average around 1,300 cubic centimeters.)

Little is known of the lifestyle of *Australopithecus afarensis* other than that it probably lived in open grasslands dotted with trees, and near rivers or lakes that supplied water and richer vegetation. There is no evidence that the species made tools. They probably ate a wide variety of plants, eggs, insects, and perhaps small amounts of meat—a dietary pattern only slightly different from that of a chimpanzee. There is also evidence that these hominids lived in family groups. Johanson has found a remarkable concentration of thirteen of their skeletons in one spot. In this cluster there were males, females, and children of all ages. It looks as if all were killed at once, perhaps by a flash flood that then quickly buried their bodies in mud before scavengers could scatter them.

By about 2.5 million years ago, *A. afarensis* appears to have evolved into at least two later forms of *Australopithecus*—a robustly built form that eventually became extinct, and a more lightly boned form that evolved onward to become modern human beings. In different regions of Africa there appear to have been different forms of both the robust and lighter forms.

AUSTRALOPITHECUS AFRICANUS and *BOISEI*— The period between 2.5 and 1.3 million years ago is one of the most confusing in human evolution because there appear to have been at least two, probably three, and possibly four different kinds of hominids in Africa. All were closely related members of the genus *Australopithecus*, but one is sometimes considered as having been advanced enough to warrant designation as the earliest member of our own genus, *Homo*.

Most experts recognize two new types of *Australopithecus* as successors to *Australopithecus afarensis*—a smaller, lighter-boned form standing four-and-a-half feet tall and spoken of as "gracile," and a larger, more massively boned form called "robust" that was pushing five feet. The gracile form is called *Australopithecus africanus* and the robust form, of which there may have been two variants, is called *Australopithecus robustus* or *Australopithecus boisei.*

Most authorities now discount the robust forms as our ancestors; they appear to have been evolutionary dead ends. Their massive jaws and teeth (molars nearly an inch across) were adaptations for processing large amounts of coarse vegetable matter. The gracile form, on the other hand, is considered ancestral by many anthropologists. Since the discovery of *A. afarensis,* however, some authorities have suggested that the gracile form was also a dead end. In this view *A. afarensis* led directly to the earliest forms of *Homo.*

Although no tools are definitely associated with any *Australopithecus* remains, authorities believe that since these hominids had brains slightly larger than those of chimpanzees (450 cc compared with 400 cc), they were probably capable of using sticks and stones with a bit more skill and purpose than do chimps. It is most likely that the various forms of *Australopithecus* lived in family groups on the African plains—sticking close to rivers or lakes and gathering a wide variety of plant foods, scavenging lion kills for meat, and perhaps catching and eating small game such as rabbits and turtles from time to time.

HOMO HABILIS—The oldest known hominid to be granted the designation *Homo,* implying that it was a true human being, is *Homo habilis.* Although in form these creatures were much like their gracile *Australopithecus* contemporaries, they were slightly taller, approaching five feet, and, more importantly, they had bigger brains— ranging from 500 cc to 750 cc. The oldest and best-known example is the famous "1470" skull found in northern Kenya by Bernard Ngeneo, who worked for Richard Leakey, a Kenyan anthropologist. It is probably between 1.8 and 2.0 million years old. (The number 1470 was its entry in the catalog of the Kenya National Museums. Almost every fossil found is given a number. The numbers also give anthropologists a convenient way of referring to fossil specimens when they talk.)

Significantly, the advent of *Homo habilis* coincides with the earliest known stone tools. These are irregularly shaped chunks of rock. Sometimes, it appears, two stones were simply bashed together, and the resulting fragments with the sharpest edges were selected for use. Sometimes stone tools were more deliberately made: A smoothly rounded cobble was struck from one side so as to detach a chunk of stone, and then struck from the other side to produce a converging facet. The intersection of the two newly exposed facets was a sharp edge on an otherwise smooth and easily held stone. Such tools are known as choppers, and experiments have shown they are fairly effective as wood-cutters or even butchering knives.

It is also significant that many of the stone tools of *Homo habilis* were found concentrated on ancient land surfaces along with animal bones that often show cut marks exactly in the places where a knowledgeable modern human would cut to get the most meat off the bone. Many of the limb bones appear deliberately smashed, as if to get at the marrow. These tool and bone sites have been interpreted as "home bases" and as butcheries. They show no evidence of fire. The home base interpretation is controversial because some authorities say that it would be possible for non-hominid forces, such as fast-flowing streams, to gather bones and stone tools into a concentration.

Because of the greater occurrence of animal bones at *Homo habilis* sites, anthropologists feel that this is the stage of human evolution when hunting began to develop as an important activity, although plants and other gathered foods such as eggs and insects undoubtedly remained staples, as they are in modern hunter-gatherer societies. At some *Homo habilis* living sites, the existence of large-animal bones suggests an improving ability to hunt as a group. Success at this would require cooperation, an important element in the increasingly complex social interactions of early people.

Since hunting big game requires venturing far from water sources, and no primate can go long without water, many anthropologists assume that *Homo habilis* invented vessels for carrying water, perhaps hollow gourds. Wide-ranging forays for plant foods probably also required the invention of baskets and other containers if any large quantities of food were to be brought back to the home base. For many reasons, including the larger size of *Homo habilis* males as compared with females, it is assumed that the males hunted while the smaller females, encumbered with children, gathered.

HOMO ERECTUS—By about 1.5 million years ago, the brain size of early hominids had increased to an average of more than 800 cc—double that of chimpanzees and more than half that of modern people. As the brain case expanded, it bulged forward and upward, and the jaws receded, creating a flatter facial profile, though with much more prominently bulging eyebrows than were known before or after. The differences were deemed to warrant a new classification—*Homo erectus*.

Over the next million years, until perhaps 400,000 years ago, the brain capacity of *Homo erectus* continued to increase, with some brains reaching 1,300 cc—the modern average. The growth, as mentioned, was concentrated in the frontal regions. In living humans these regions are important for muscle coordination and the ability to concentrate on complex tasks and plan ahead. The brain expansion led to more complex behaviors which, in turn, created new opportunities for further brain development to prove advantageous, both in social organization and in survival skills. Several discoveries attest to a growing mental ability. For example, *Homo erectus* was the first hominid to expand out of Africa and into Eurasia, a move that required an ability to cope with a wide variety of climates and habitats. These early people ranged as far north as Beijing, (formerly spelled Peking), where the winters are cold. *Homo erectus* was also the first hominid to use fire, probably for warmth and cooking. Nobody knows whether it wore clothes, but it seems unlikely the species could have survived northern winters without wraps made of animal skins as minimal clothing.

Besides these features, *Homo erectus* developed many new stone tools. Instead of randomly broken rocks, these early people made more standardized forms including superior hunting weapons and butchering tools, reflecting a growing reliance on meat as success at hunting apparently became more assured. The most extraordinary stone tool was the hand ax—a beautifully shaped object resembling a flattened tear drop. Its shape, pleasing to modern eyes and, in a sense, more beautiful than it needs to be, suggests a fairly developed esthetic sense. Hand axes have been found all over the *Homo erectus* world in thousands of examples, all reflecting the same basic design. Although nobody today really knows what these hand axes were used for, they are powerful evidence of an ability to conceive a design and then work a piece of stone until it was achieved. Many linguists feel the ability to make hand axes required a kind of conceptualizing or symbolizing ability much like that needed for language. *Homo erectus* may well have had at least a rudimentary spoken language.

The spread of this species into widely separated parts of the world also had an effect on subsequent human evolution. Since environmental conditions varied from place to place, local evolution produced the *Homo erectus* equivalent of different races. These differences, however, should not be confused with racial differences in modern human beings, who did not appear until some 300,000 years after the last *Homo erectus* was gone.

EARLY *HOMO SAPIENS*—The past 500,000 years— the period since the latest clear-cut examples of *Homo erectus* lived—have seen no sudden transformations of one kind of hominid into another. The late forms of *Homo erectus* appear to have evolved gradually into *Homo sapiens*, our own species. In fact, many specimens from Africa, Asia, and Europe show such a mixture of traits that anthropologists debate whether they should be assigned to *erectus* or *sapiens*.

By about 400,000 years ago, however, continued flattening of the face, steady increases in brain size, and major
technological and conceptual advances give evidence of
people recognizably like ourselves. However, they are not
completely like members of our subspecies, *Homo sapiens
sapiens*. Their brains, on average, were still only about 83
percent the size of ours, and their faces had bulging brow
ridges.

The early members of our species were better hunters
than *Homo erectus*, bringing down deer, elephant, boar,
ibex, rhinoceros, and other large animals. They cooked
their meat, wore clothes of animal hide (at least in cooler
climates), made wooden tools (including spears with fire-
hardened points), and constructed huts of animal skins
stretched over wooden poles. Some of the largest of these
were ovals nearly fifty feet long, with walls consisting of
three-inch diameter stakes driven into the ground and
buttressed with rocks around the outside. At the center
was a hearth, probably vented through a hole in the roof.

A major technological advance of the early sapiens was
the invention of the "prepared-core" technique of making
stone tools, a method that appeared in Africa and Europe.
The old method had been to chip a rock, hoping to come
closer and closer to the desired shape. The rock itself told
you whether you were succeeding. The prepared-core
method requires another level of conceptual distance between the goal and the work done to achieve it, because
the first stage of work must produce something that does
not look at all like a tool. First the craftsman chips a flint
nodule into a kind of cylindrical loaf shape with a flat bottom, domed top, and vertical sides. From this core one
can then detach, with a single blow, long flakes that make
deadly knives. Older methods of making knives involved
repeated chipping to fashion a long, sharp edge. The result was a ragged knife edge. The new method yielded
smooth edges. To hit upon the prepared core method,
somebody had to have been trying to find a better way of
making a tool—an intellectual stance appropriate to our
own time.

The technology of early forms of *Homo sapiens* points
to a mental sophistication sufficient for spoken languages
approaching ours in usefulness. There are other clues that
early sapiens were using mental processes like our own.
At one 400,000-year-old site excavators found lumps of
red ochre. Its only conceivable use was as a pigment for
coloring things. More tantalizing is a 300,000-year-old ox
rib upon which were engraved sets of parallel lines. Some
zig-zag, some meandering, some standing alone. They look
as if they were supposed to mean something, though it is
not clear what.

HOMO SAPIENS NEANDERTHALENSIS—The
Neandertals (the modern spelling of their colloquial name,
though not of the scientific name, has been changed to reflect the correct pronunciation) are undoubtedly the most
misunderstood people of all time. The conventional image
of them is of hulking, dull-witted brutes, far removed
from our own cohorts. This widespread impression of the
Neandertals dates back to 1856, when the first find of a
Neandertal skeleton occurred. It was the first extinct
form of human ever recognized, and it happened (we know
now) to have been an unusually robust individual. In eyes
that had never seen more primitive hominids, the "Neandertal man" was considered as about halfway between
apes and humans.

All this has changed. Today most authorities recognize
that the Neandertals were, in fact, members of our own
species. They appeared between 100,000 and 70,000 years
ago and are now classified as *Homo sapiens neanderthalensis*. Their brains, equal to ours in size, represented a clear advance over that of the early sapiens, and
may have been virtually identical to our own. Some authorities dissent on this, however, maintaining that there
were significant mental differences.

Many specimens are known today from all parts of Europe and western Asia, and it is clear they were a highly
variable people. Some, called classic Neandertals, were
massively-boned, heavy-browed, large-nosed people much
like the 1856 specimen. But many others, including some

mixed into the same populations from which the classic examples came, show a wide variety of skeletal features, including every trait familiar in living Europeans. They were, however, more heavily muscled than are most people today.

The Neandertals possessed all the technological skills of their early sapient ancestors. They developed variations of the prepared-core technique to make a wide variety of fairly standardized stone tools, including some of the first long flint knives. Perhaps the most significant evidence of the Neandertals way of life is in the way they dealt with death. Instead of simply abandoning their dead, as older hominids probably did, the Neandertals conducted funerals—ritual burials complete with grave goods including tools and food. Some graves are surrounded by ibex skulls. One grave, in Iraq, even had clusters of pollen grains that have been identified as the remains of flowers, all but one from species that are known today to have medicinal properties. Far from being brutes, the Neandertals give us our earliest evidence of apparently spiritual qualities in human beings, including a respect for the individual, concern over his passing, and perhaps even a religious system involving belief in an afterlife.

The greatest remaining controversy over the Neandertals is whether they evolved into fully modern people or were driven to extinction by an invasion of modern peoples who migrated from elsewhere. Excavations of Neandertal sites suggest a sudden replacement by fully modern people between 30,000 and 40,000 years ago. But the evidence is not completely clear and the case is still open. While the Neandertals occupied Europe, peoples at least as advanced were living in Africa and Asia. Some experts see these peoples as racial variants of the Neandertal. Very little is known of their culture.

HOMO SAPIENS SAPIENS—People anatomically indistinguishable from ourselves were living in Africa, Asia, and Australia by about 40,000 years ago. They were not yet in Europe, which was already occupied by the Neandertals. In the next few millennia, however, the Neandertals would disappear from Europe, succeeded by fully modern peoples. Whether the modern peoples exterminated the Neandertals or interbred with them is a major controversy of anthropology.

Nobody knows where modern people first appeared. Recently, however, there has come tantalizing evidence that it may have been in southern Africa at least 115,000 years ago—before the Neandertals appeared in Europe. The date on the African skulls, however, is not certain, and they could be modern skulls of more recent age.

Yet even if we accept a date as late as 40,000 years ago for the beginning of *Homo sapiens sapiens*, some startling conclusions result. This would mean, for example, that human beings with all the intelligence of modern humans existed for at least 25,000 years as hunters and gatherers, making their living much as did their hominid ancestors for the previous two or three million years. Animals were not domesticated until perhaps 18,000 years ago, and plants not until 12,000 years ago. The earliest permanent settlements appeared 9,000 years ago and the first real cities only 5,000 years ago. Civilization based on agriculture—like ours today—is only a fairly recent development in the existence of fully modern human beings.

The early moderns produced exquisite stone tools, including knives, spearpoints, and arrowheads as deadly as any made of steel. They invented the bow and arrow and the spear-thrower, a device that gives added leverage to propel a spear farther and harder. These two inventions made it possible to kill dangerous prey without getting too close. Fewer hunters were lost and more meat came home. The early modern peoples also developed hooks and nets for fishing, adding abundant new sources of protein and variety to their diets.

The prehistoric moderns apparently enjoyed such material surpluses that they could afford to permit individuals not to be productively employed at all, but instead to become consummately skilled artists. From the beginning of our subspecies, we produced high-quality sculptures such as the famous Venus figurines, which may have been icons of an early religion. Between 20,000 and 15,000 years ago we began producing the great cave paintings of southern France and northern Spain. The art is of a highly sophisticated style and technique, requiring a mastery of skills possible only in a full-time "professional."

Such were the achievements of "savages." But, despite stereotypes, hunting and gathering is not a burdensome way to live. Modern hunter-gatherers, confined nowadays to the poorest land, meet all their material needs by working only four or five hours a day. Imagine, then, how abundant life must have been in the more luxurious habitats available 25,000 years ago. The wonder then is not that it took us so long to reach civilization, but why we ever abandoned hunting and gathering at all.

See CHROMOSOMES; GENES; HOMINID; EVOLUTION.

HUTTON, J., 1726–1797, Scottish geologist. See DARWIN; UNIFORMITARIANISM.

HUXLEY, JULIAN, 1887–1975, British biologist. See MODERN SYNTHESIS.

HUXLEY, THOMAS HENRY, 1825–1895, British biologist. See DARWIN.

HUYGENS, CHRISTIAAN, 1629–1695, Dutch physicist and astronomer. See NEWTON.

HYBRID—Any time you crossbreed two different varieties of plant or animal (normally both are of the same species), you get a hybrid—an offspring that combines many of the hereditary traits of both varieties. Very often the hybrid organism, be it plant or animal, grows much more vigorously than its purebred counterpart—a phenomenon known as hybrid vigor. Hybridization is often used in agriculture to produce superior types of plants or animals.

Generally speaking, the parent stocks of hybrids represent two relatively purebred varieties. In other words, both parents have such nearly identical sets of genes that their offspring turn out to be very close in traits to their parents. Hybrids, on the other hand, do not "breed true." Cross two hybrids (even two from the same parents) and the offspring will show an assortment of features ranging from those of the male parent to those of the female parent. The desirable combination is unlikely to come up again. This is why it is not recommended to save seeds from hybrid vegetables.

Some hybrids are produced by crossbreeding two different species. The classic example is the horse-donkey cross that produces a mule. The mule illustrates the extreme problem in reproducing hybrids—that two parental sets of genes that can cooperate to produce a healthy individual may, nonetheless, be too unlike one another to produce viable sperm or eggs. Hybrids of different species, when they can be produced at all, are almost always sterile. Hybrids produced within a species, however, are fertile.

See SPECIES.

HYBRIDOMA—One of the most powerful new technologies developed by biologists in recent years is an unusual form of life that grows only in the laboratory. It is a mass of human or animal tissue that will live and grow indefinitely in a jar as long as technicians supply liquid nutrients to it. As it lives, the tissue mass produces huge quantities of one specific kind of antibody, a protein molecule that is becoming a workhorse of medical diagnosis and treatment.

Each cell of the tissue mass, which is called a hybridoma, is a hybrid of two different kinds of cell—a cancer cell and another kind of cell, called a B lymphocyte, that normally makes antibodies. The two cells are put together and dosed with certain chemicals, their cell membranes dissolve, and they join to become one cell. After one or two cell divisions, the two nuclei, each carrying the genes of its original cell, merge into one nucleus. The lymphocyte acquires the cancer cell's immortality, its ability to keep on dividing forever, as opposed to non-cancer cells, which die after about fifty cell divisions. The fused cell grows into what amounts to an antibody-making tumor.

Since each type of B lymphocyte used in hybridoma technology produces only one kind of antibody, the hybridoma churns out huge quantities of these useful molecules. Moreover, the antibodies produced are all of one kind, made by the descendants—or a natural clone—of the same cell. These monoclonal antibodies, as they are called, are used to diagnose and treat many kinds of disease.

See IMMUNE RESPONSE; ANTIBODY.

HYPOTHESIS—See THEORY.

IMMUNE RESPONSE—The response of the body's immune system to invasion by some foreign protein molecules, which are most often protein molecules on the outer surfaces of bacteria and viruses, although they may also be the proteins on a transplanted organ. Contrary to the way in which its name strikes some people, the immune system is not a system that is immune. Perhaps a better name would be the immunological system. The immune response involves many things, including the manufacture of antibodies—special protein molecules tailored to bind to a foreign protein. The antibodies either directly render the invader innocuous, or immobilize it until another element of the immune system—special white blood cells—can come along and literally eat it.

Since it takes the immune system some time to mount a counterattack on invading microbes, vaccines are used to fool the body into thinking an attack is underway. The microbe in the vaccine cannot reproduce, but its proteins still prompt the immune system to make the appropriate antibodies. These continue to circulate in the blood for years, ready to quickly put down an invasion before it can produce serious disease.

See JENNER (who devised the first vaccine); PASTEUR (who developed it much further).

INTERFERENCE—In most cases this term refers to a phenomenon that occurs when light waves traveling together in a parallel path interact. If the waves of a given color are out of phase by exactly one half of a wavelength, they cancel one another out, and the color produced by that wavelength can no longer be seen. This is called destructive interference. On the other hand, if the waves of a particular color are superimposed on one another in phase, the result appears as an unusually brilliant display of the color. This is constructive interference. Another way of describing destructive interference is to say that at the point where one light wave is at its peak of its energy, the interfering light wave is at its lowest point, which actually has a negative value. When these interfering waves reach the eye, the negative value of one wave exactly cancels out the positive value of the other. If the two waves are in phase—as in constructive interference—the two peaks arrive simultaneously and their value is added. The result of constructive interference is typically known as iridescence, and can be seen in the colors of soap bubbles, peacock feathers, pearls, and other natural objects.

INTERFERON—When animal or human cells are infected by a virus, they often produce special proteins that diffuse into neighboring cells. The cells that receive these proteins become temporarily resistant to the virus because the protein interferes with the virus's attempt to reproduce inside the cell. When researchers discovered the existence of these proteins, they named them interferons (there are several kinds), and hoped that it might be possible to use them in medicine to help people resist or overcome viral infections. Some types of interferon were especially attractive for this because they seemed to block reproduction of many kinds of viruses. There was also hope that interferons might be useful against cancer.

Early tests of interferons, begun with wildly exaggerated publicity, proved generally disappointing. For one thing, their quantities were extremely limited because they were harvested from animal cells grown in the laboratory. Now, through recombinant DNA technology, larger and more pure supplies are being developed and tested. They may yet prove highly useful.

INTRON—A portion of a DNA sequence that contains genetic gibberish. The term is used to contrast this part of a DNA sequence with other parts known as exons that encode proper genetic messages. Many genes contain intervening intron sequences.

See NONSENSE DNA; GENES, HOW THEY WORK; GENETIC CODE; EXON.

ION—An atom that has an electric charge, either positive or negative. All atoms are fundamentally neutral in terms of their electric charge. This is because the number of negatively charged electrons orbiting the nucleus equals the number of positively charged protons in the nucleus. However, many atoms easily gain or lose electrons. An extra electron gives the atom a negative charge, while a loss of one electron leaves the atom with a net positive charge. Some atoms may gain or lose more than one electron, producing more strongly charged ions.

The ionic forms of some kinds of atoms are common in nature. In such cases the ions are almost always bound to other ions with opposite charges. The bond is of a special, relatively weak type, called an ionic bond.

See ATOM; CHEMICAL BONDING, NATURE OF.

IONIC BONDING—See CHEMICAL BONDING.

ISOTOPE—Atoms of the same element are isotopes of one another if they have different numbers of neutrons in their nuclei. Since the fundamental chemical nature of an atom is determined by the number of protons in the nucleus and the number of electrons orbiting the nucleus (and not by the number of neutrons), all isotopes of an atom are capable of the same chemical reactions. Those with extra neutrons, however, are heavier, and in some cases are radioactive; that is, they may spontaneously emit a burst of energy and change themselves into a more stable form.

Most elements occur naturally in the form of several isotopes, although one is much more abundant than the others. Carbon, for example, is 98.9 percent composed of the isotope C^{12}—which has 6 protons and 6 neutrons in its nucleus—and only 1.1 percent composed of C^{13}, which also has 6 protons but has 7 neutrons instead of 6. C^{14} is vastly rarer and is produced when cosmic rays strike nitrogen atoms in the upper atmosphere, converting them into a radioactive form of carbon.

Like C^{14}, many artificially made isotopes are radioactive and, therefore, useful in research. For example, a drug can be made with a radioactive isotope of one of the constituent atoms instead of the usual atom. The isotope fulfills the chemical role of the atom in the drug, but will eventually emit a tiny burst of energy that can be detected with instruments. For this reason, such a molecule is said to be radioactively labeled. By scanning the body with radiation detectors, researchers can then tell where the drug has accumulated. The amount of radioactivity that is released is usually well within the limits accepted as safe.

See ATOMIC STRUCTURE

JACOB, FRANCOIS, 1920–, French bacteriologist. See GENES, HOW THEY WORK.

JENNER, EDWARD, 1749–1823, English physician— Edward Jenner developed the world's first vaccine, which when inoculated prevented smallpox. He thus launched the field of immunology and devised one of medicine's most effective strategies for dealing with disease. Some 180 years after Jenner administered his first dose of smallpox vaccine, the disease was utterly eradicated from the Earth. Smallpox is the first disease to be wiped out completely and, for various reasons, it may be the only one likely to reach this state.

Jenner was the son of an English country clergyman who died when the boy was only five. He grew up in the country with a keen interest in nature that persisted into adult life. After apprenticeship to a prominent London surgeon, Jenner returned to the country as a doctor. He divided his time between seeing patients and observing wildlife behavior. He found, for example, that it was the newly hatched cuckoo, not the parent cuckoo as had been believed, who pushed the other eggs from the nest. For such observations, Jenner was elected a fellow of the Royal Society.

His more important contribution to science came from his dissatisfaction with the prevailing method of trying to prevent smallpox, then a common cause of death and disfigurement in England. The method was to take "matter" from the sores of a victim of the milder form of smallpox and put it in a scratch on the skin of the person being immunized. It was already known that survivors of a smallpox infection were immune to subsequent attacks, and that surviving the mild form of the disease protected one against the severe, often fatal form. The problem was that the mild form sometimes progressed to the severe form. All too often the person seeking immunity was killed by the deliberate infection. It was also known that people who had been infected by cowpox, a relatively mild ailment contracted from infected cows, seemed immune to smallpox.

Realizing that humans almost never died of cowpox, Jenner innovated by trying to inoculate people with cowpox "matter." In 1796 Jenner found a young dairymaid with fresh cowpox sores on her fingers. He scraped material from the sores and used it to inoculate an eight-year-old boy. The boy developed a slight fever and a minor lesion. Six weeks later Jenner inoculated the boy with smallpox matter. There was no reaction. The boy was immune.

Jenner promptly sent a report to the Royal Society, for publication, but it was refused. He repeated the experiment with twenty-two more patients, always with success. Then, in 1798, he independently published a small book on the experiments. The reaction was not immediately favorable, and Jenner had difficulty getting more volunteers for his studies. Other doctors siezed upon the discovery and tried to claim it as their own. Within a few years the death rate from smallpox plummeted. Eventually Jenner gained recognition and won not only worldwide acclaim but parliamentary grants totaling 30,000 pounds.

Although Jenner did not fully understand the mechanism responsible for immunity, he introduced the term "virus" to describe the "matter" producing infection, and laid the foundations for the development of the modern science of virology.

See GERM THEORY; IMMUNE RESPONSE; VIRUS.

JOHANSON, DONALD, 1943– , American paleoanthropologist. See HUMAN EVOLUTION.

JOULE, JAMES P., 1818–1889, English physicist. See THERMODYNAMICS.

KANT, IMMANUEL, 1724–1804, German philosopher. See HELMHOLTZ.

KELVIN, WILLIAM THOMSON, LORD,— 1824–1907, British physicist—William Thomson, who was named Lord Kelvin in 1892, is remembered most often today for the temperature scale he devised (in which temperatures are measured in degrees Kelvin, with zero being absolute zero—the complete absence of heat), but his major contribution was in formulating the law of conservation of energy. This is the idea that any form of energy—be it heat, light, motion, or another—can be converted into another of the forms. Furthermore, though later modified by Einstein's theory of relativity—which shows that matter and energy are alternative forms of the same thing—Thomson's law states that energy can neither be created nor destroyed, but only converted from one form to another. If this were not true (apart from relativistic exceptions), no motor could work nor could life exist.

Although the law of conservation of energy was Thomson's chief scientific contribution, he also made major advances in engineering, especially with various inventions such as telegraph receivers and electricity detectors that were used in the early trans-Atlantic telegraph cables. As a result of his inventions he earned vast patent royalties and became quite wealthy. With his money he even bought a 126-ton sailing yacht and traveled the world for pleasure.

Thomson was born in Belfast. His mother died when he was six. His father wrote textbooks and, after moving to Scotland, taught mathematics at the University of Glasgow. As a small boy, Thomson learned the latest advances in mathematics at his father's knee even before they were being taught in colleges. It is said that because Thompson was a very submissive son and his father a domineering man, the two developed an unusually close relationship that helped develop the boy's extraordinary mind.

Young Thomson entered the university at Glasgow at the age of ten. At the age of fifteen he had won a major award for his scientific work on the shape of the Earth. Then, in 1846, when a professorship at Glasgow fell vacant, the elder Thomson pressured the university to name his son to the post. He succeeded and, at the age of twenty-two, Thomson was unanimously elected to the chair of natural philosophy.

Thomson developed his scale of absolute temperatures as a more convenient method of measuring temperature when doing research on the relationship between heat and work. (In the practical world, this relationship is mediated by engines which are, at heart, devices for transforming heat energy into work.) The centigrade scale in use at the time (called the Celsius scale today) used the freezing/melting point of water as zero degrees. This, as in the Fahrenheit scale, produced the odd situation of forcing one to use negative numbers for temperatures that—although below zero—still indicated a positive quantity of heat. Ice at 10 degrees below zero is warmer than ice at

20 below. Thomson devised his scale with zero as the absolute coldest that anything can be. As it happens, his absolute zero is 273 degrees below the freezing point of water. This means that an ice cube at freezing temperature still has in it more than twice as much heat as was removed in cooling the water from the boiling point down to the freezing point. The size of the degree on Thomson's scale is the same as that on the centigrade scale, which is almost twice the size of a Fahrenheit degree. On the Fahrenheit scale, absolute zero is nearly 500 degrees below the freezing point of water.

Once he had a sensible way of measuring temperatures, Thomson set about a mathematical treatment of the relationship between heat and mechanical energy, or work. It led to his formulation on the conservation of energy.

Thomson was not always a success in science. Once when he sought to apply his equations to the heat emanating from the sun and the Earth, he came to the conclusion that only a mere one million years ago, both had been too hot for life to exist as we know it. He argued that Darwin must have been wrong in suggesting that life has been thriving on the Earth for uncounted millions of years.

Though Thomson would be proven wrong, it was not obviously so in his day, and his fame as a theoretical scientist with some 600 published papers to his credit, and a practical engineer with dozens of patents, was vast. At the age of sixty-eight he was named Lord Kelvin. He received so many other awards and honors that, it was said, he was entitled to more letters after his name than any other man in Britain. Kelvin died at the age of eighty-three and was buried in Westminster Abbey.

See CONSERVATION OF ENERGY; THERMODYNAMICS.

KEPLER, JOHANNES, 1571–1630, German astronomer, physicist—Johannes Kepler ardently believed in the century-old theory of Copernicus that the Earth orbits the sun, but he could not make astronomical observations fit Copernicus's insistence that the planets trace perfect circles about the sun. He found, instead, that their orbits were elliptical, or egg shaped. This became the first of Kepler's three laws of planetary motion. For a time Kepler worked with the eccentric but meticulous Tycho Brahe, improving the accuracy of astronomical data to a point that demanded improvements in cosmological theory. Kepler and Galileo were contemporaries and corresponded. At a time when Galileo's ideas on Copernicanism were being ridiculed, Kepler accepted and acknowledged them in his own writings. Galileo, on the other hand, seems to have ignored Kepler's work in his scientific writings. Although both advocated the Copernican theory, it was Galileo's additional astronomical observations, along with his greater eloquence and more polemical nature, that won him the greater position in history.

Kepler's life was not a happy one. He was the premature child of an innkeeper's daughter and a mercenary soldier who seems to have abandoned the family early on. Small and in chronic poor health, Kepler might never even have gone to school had it not been for the policy of the dukes of Wurttemberg to award scholarships to the brightest sons of the poor. In college, Kepler's astronomy professor was one of the few who accepted Copernicanism and taught it.

Thus Kepler began his interest in astronomy as one of the few Copernicans of his time. Kepler, however, was also a believer in the intrinsic harmony of the universe, a concept that to him was bound up with Platonic notions of ideal forms. His first attempt to explain the spacing of the planets created the famous but wholly unfounded notion that between the spheres, or "shells" in which the first six planets had their orbits (and whose equators were the planets' circular orbits), one could fit huge versions of the five Platonic solids (octahedron, icosahedron, dodecahedron, tetrahedron, and cube). As it happens, presumably coincidentally, the scheme worked—at least within the poor accuracy with which the planetary orbits were known. When Kepler improved the accuracy of the orbital measurements (reducing the error to less than 1 percent of what it had been), he had to confront the truth that the orbits were not circular after all.

Kepler abandoned his Platonic idealism but, in a sense, compensated for it by discovering two other "laws" that helped to retrieve his sense that the cosmos must be harmonious. One—his second law of planetary motion—states that if you draw a line from the sun to a planet, the line will—as the planet orbits the sun—sweep equal areas of space in equal times. In other words, when a planet is moving through one of the ends of its ellipse, the line to the sun will be longer but the planet will be moving more slowly. As the planet swings back closer to the sun, the line will get shorter but the area it sweeps in any period of time will stay the same because the planet speeds up. Kepler's third law came even closer to the mathematical harmony he sought to find. It said that if you cubed the average distance between the sun and any planet, and if you squared the time it took that same planet to complete its orbit, the two resulting numbers would always have the same ratio, no matter which planet you were concerned with. Kepler's laws laid the foundation for Isaac Newton to come along, half a century later, and formulate his theory of universal gravitation.

Kepler produced his works in the face of great personal adversity. His scholarships had put him through school and he had worked for a few years as a teacher before deciding in 1600 to go to Prague and join forces with Tycho Brahe, who was then imperial mathematician to the Emperor Rudolph of the Holy Roman Empire. Less than a year later Tycho died and Kepler succeeded him. Everything then went well for about ten years, after which disaster struck. The Thirty Years War rolled into Prague, and Rudolph was forced to abdicate. In the same year that Kepler's wife died, his children were stricken with smallpox and one of them died too. The next year Rudolph died. Kepler was reappointed imperial mathematician, but he left Prague and lived the next fourteen years of his life in Linz, Austria. Then his aged mother, back in Wurttemberg, was indicted for witchcraft, and Kepler went to defend her in a trial that dragged on for three years. Further troubles, including the failure of his patron to pay, kept Kepler in near poverty until his death.

See COPERNICUS; GALILEO; NEWTON; BRAHE.

KOCH, ROBERT, 1843–1910, German bacteriologist— Robert Koch, pronounced "coke," was the first person to prove that germs cause disease. In the decades preceeding Koch's work, others had suspected as much, and bacteria had been discovered, though not linked to disease. But it was Koch, a medical officer in the German government, who in 1876 actually found the first known organism that caused a disease. It was a type of bacteria responsible for anthrax, a disease of hoofed animals. A few years later, Koch made a more famous discovery, finding the bacteria that cause tuberculosis, then the most feared of all human diseases. Until Koch, scientists could not fully shake off the ancient idea that diseases were caused by bad air, or miasmas, arising from swamps or refuse heaps. In fact, the name "malaria" means bad air. After Koch, it came to be recognized that the world of living creatures included some very small organisms whose habitat was the human body.

Koch, the son of a mining official, was born in what is now West Germany. He became a doctor and was appointed district medical officer, a relatively humble position in which, with only modest resources, he set up a small laboratory and performed the experiments that would eventually earn him a Nobel prize and the honor of being considered the founder of bacteriology.

To make his discoveries, Koch had to invent many of the techniques for collecting infected matter from patients, growing the bacteria in this matter in the laboratory, and studying these bacteria under the microscope. One of his techniques remains fundamental to bacteriology and medicine today, and is used, for example, when doctors take a "throat culture." This involves smearing some of the presumably infected matter on a layer of nutrient in a dish. If bacteria are present, they soon grow into visible clumps, or colonies, on the nutrient. Koch also found one of the nutrients that is still widely used—agar, a gelatinous substance obtained from seaweed and into which other nutrients may be mixed.

Although Koch deserves the credit for establishing the germ theory, Louis Pasteur, the French chemist and microbiologist, played a role in its establishment and is sometimes erroneously given full credit for it. A decade before Koch proved the germ theory, Pasteur had traced two diseases of the silkworm to microbes, and had found ways to prevent their spread, thus saving the French silk industry. Still, Pasteur's evidence was circumstantial, and he did not prove his case as rigorously as Koch would. Moreover, few scientists who accepted the silkworm findings generalized them to human diseases.

Koch's rigor is best demonstrated in the three rules he formulated to prove that any given disease is caused by a specific organism—rules that medical scientists use to this day. The rules are known as Koch's postulates, and they stipulate that the organism must be found in all cases of the disease examined, that the organism must be grown in a pure culture in the laboratory, and that the organisms grown in the laboratory must be capable of causing the disease when given to a human or animal subject. In practical terms, these rules cannot always be followed perfectly. For one thing, there are now ethical constraints against deliberately infecting people. This is why animals are usually used in testing the ability of an organism to cause infection. Unfortunately, several human diseases have no counterpart in animals.

Following his successes with anthrax and tuberculosis, Koch indulged his passion for foreign travel, and journeyed in many parts of Africa and Asia, advising on various infectious diseases and discovering the germs that cause cholera, amoebic dysentery, and Egyptian conjunctivitis. In a way, he became the first in a long line of international health experts, a tradition that has grown ever since.

See BACTERIA; GERM THEORY; PASTEUR.

LAMARCK, JEAN-BAPTISTE DE, 1744–1829, French biologist, evolutionist—Lamarck does not fully deserve the poor reputation he has today. After all, he did assert, half a century before Darwin did, that all living things evolved from simpler forms. Where he went astray, we now know in hindsight, was in thinking that acquired traits are passed on to offspring. More specifically, Lamarck thought that evolutionary change resulted from the use or disuse of specific organs. In other words, and to cite the most famous example of his doctrine, giraffes got longer necks by stretching them into treetops to eat, and passed this trait on to their offspring. By the same token, Lamarck thought that lizards that didn't use their legs much had them waste away—an evolutionary step leading to the origin of snakes.

Lamarck had no knowledge of genes. Gregor MENDEL'S pea-breeding experiments occurred after Lamarck died, and were not really known to most of the scientific world until the twentieth century. Darwin didn't have any knowledge of genes either, but he was lucky in that his proposed mechanism of evolution was compatible with the existence of genes. Long after Lamarckism had been discarded by most scientists, it enjoyed official revival in the Soviet Union because it fit well with the Marxist view that environmental factors can alter human nature. The early Soviets hoped that forced socialism would eventually cause selfishness to wither away, and that "the new Soviet man" would be innately selfless. A Soviet biologist named Trofim Lysenko advocated Lamarckism well into the twentieth century, and government endorsement of his views paralyzed Soviet agricultural progress at a time when other industrialized countries were engaged in successful plant-breeding programs based squarely on the teachings of Mendel and Darwin.

Lamarck also deserves recognition as an originator of the modern concept of the natural history museum as a repository of plant and animal specimens. Until his time most museums were haphazard collections of curiosities donated by private collectors. Shortly after the French

revolution of 1789, Lamarck persuaded the national assembly of France to create a new museum that would
have systematically organized collections of specimens
representing all the major classifications of life. Lamarck
was put in charge of the section dealing with animals that
have no bones, a category for which he coined the term
invertebrate. (Lamarck also seems to have been the first
person to use the word biology, in 1802.) Lamarck was
able to differentiate crustaceans, insects, and arachnids
(spiders) from one another—a discernment that has stood
the test of time.

See MENDEL; EVOLUTION THEORY.

LAMARCKISM OR LAMARCKIAN EVOLUTION—
A hypothetical form of evolution in which traits acquired or
developed by one generation are passed on in fully developed form to the next.

See LAMARCK; EVOLUTION THEORY.

LAPLACE, PIERRE-SIMON, 1749–1827, French
mathematician, physicist—It is not often mentioned, but
when Isaac Newton unveiled his grand theories of the
movement of the heavenly bodies, there were a couple of
little problems with them. There were certain perturbations in the movements that, Newton thought, would
gradually build up, leading eventually to the destruction
of the world. Newton left these irregularities in his theory, and stated that he was sure God would intervene in
time and set everything aright again. It was Laplace who
redid the calculations and found that while Newton was
right about the existence of the perturbations, he was
wrong about their being cumulative. The solar system
was basically stable, Laplace found.

The book in which Laplace put forth these calculations is
his greatest work, and it is said that when Napoleon, who was
an enthusiastic patron of mathematicians, read the book, he
noticed that Laplace had not once mentioned God. Laplace
proudly replied, "I have no need of that hypothesis."

In a popular version of the book, Laplace added his

speculation that the solar system began as a great whirl-ing cloud of gas, a nebula that condensed to form the sun. In the process, the planets were flung off the condensing sun. Nobody knows for certain how the plan-ets were formed, but Laplace's idea remains a contending theory.

Laplace's second greatest contribution was in the realm of probability theory, which can be more easily under-stood as the mathematics of games of chance. Through mathematical descriptions of probabalistic events, he laid the foundation for much of the way in which modern sci-ence deals with large quantities of data.

Laplace was born to a peasant farmer in Normandy, where his mathematical ability soon became apparent in school. He then went to Paris and set about trying to make his way as a mathematician. During this time, he wrote a letter to a prominent mathematician in the city and discussed his views on various principles of mechanics. The mathematician was so impressed that he recommended Laplace for a professorship at the École Militaire. From there, Laplace's fame and influence grew.

At one time Laplace worked with the chemist Lavoisier on experiments showing that respiration was a form of combustion. Laplace was also, in a sense, a survivor. When the French Revolution came in 1789, Lavoisier was beheaded but Laplace was unaffected, probably because he held no strong political views. When Napoleon took the throne, Laplace easily became a favorite of the emperor's and, when the Bourbons were restored to the monarchy, had no trouble signing the decree banishing his former pa-tron.

Along the way Laplace helped create the metric sys-tem.

See NEWTON; LAVOISIER.

LAURASIA—The ancient continent that included what is now North America.

See GONDWANA (the other ancient continent).

LAVOISIER, ANTOINE LAURENT, 1743–1794, French chemist—You can observe a phenomenon closely, and you may think you know what's going on, but if you don't make careful measurements, you're likely to be fooled. A common enough tenet of science today, this idea took a lot of establishing to catch on. Galileo introduced it to physics in the sixteenth century, but chemists didn't catch on until Lavoisier preached it in the eighteenth century. He also practiced it, and in the process demolished the curious doctrine of phlogiston—the mysterious substance that was thought to be liberated by all objects when they burned. (Among the problems with phlogiston was that to explain all forms of combustion, this invisible form of matter sometimes had to have weight, sometimes had to have no weight, and sometimes had to have negative weight. Lavoisier, as we shall see, discovered what really happened when things burned, and turned phlogiston into one of the quaint notions of early science that chemistry students of today, lacking a historical sense, like to ridicule.)

Lavoisier also established the law of the conservation of mass—the idea that mass was neither gained nor lost in burning or other transformations, but merely changed from one form to another. And he established the first standardized nomenclature for chemical compounds, overthrowing the alchemists' tradition of using such strange and obscure names that it was practically impossible to tell what they were talking about. As a further contribution, Lavoisier wrote the first modern textbook of chemistry. For all these achievements, he is regarded as the founder of modern chemistry. Benjamin Franklin and Thomas Jefferson were frequent and admiring visitors to his Paris laboratory.

Thus it may seem strange that this champion of rationalism and science was condemned to death by the new French Revolution, guillotined in 1794, and his body dumped in a mass grave.

Right as he was about so many things chemical, Lavoisier possessed some traits that did not endear him to the French citizens of the time nor to other scientists. Born to a wealthy Paris family (his father later bought him a noble title), Lavoisier had the freedom to pursue intellectual interests such as science, and money to invest. He put his money into the Ferme Générale, a private company empowered by the monarchy to collect taxes. The company paid the government a fixed fee and kept any "taxes" it could collect beyond the fee itself. Its profit margin, in other words, lay in gouging the taxpayers. Lavoisier, like other investors in the Ferme Générale, was considered a "tax farmer" and was much resented by the French populace. Nonetheless, profits from tax farming enabled Lavoisier to set up the lavishly equipped private chemical laboratory from which his many contributions to science emerged.

Lavoisier's fate at the hands of the revolutionaries might have been prompted by tax farming alone, but the scientist ensured it through another circumstance. As a member of the French Academy of Sciences, Lavoisier blackballed the application of one Jean-Paul Marat, a journalist who dabbled in science. Marat's scientific work, on the nature of fire, was badly done, Lavoisier felt, and unworthy of the honor implied by academy membership. Marat, of course, went on, some twenty years later, to become a leader of the revolution and a zealous prosecutor of the tax farmers. Having nursed his grudge long enough, Marat ordered Lavoisier arrested, tried (Lavoisier denied, falsely, that he was a tax farmer), and executed.

Lavoisier's contributions to science, however, still stand as monuments to the scientific method as it is practiced today. To appreciate them, it is necessary to realize just how wrong chemists could be in those days when medieval alchemy still held sway. It was believed, for example, that matter could be transmuted, not merely in the form of base metals into gold, but even of water into earth. Lavoisier's experiments on this latter point illustrate his method. It was known, for example, that if you heated water long enough, sediment would appear in the bottom of the vessel. Lavoisier examined this phenomenon with careful control and measurement. His flask had an attached condenser to catch the steam, cool it back into liquid water, and let the water drip back into the flask. He weighed both the flask and water and then boiled the water for 101 days. Sure enough, sediment formed in the bottom of the flask.

To establish where it came from, Lavoisier again weighed the flask and the water, as well as the sediment. The water had not changed weight, but the flask was lighter by an amount exactly equal to the weight of the sediment. Heat, Lavoisier proved, did not cause water to transmute into "earth." Instead, the hot water had leached minerals from the glass.

Lavoisier's work with heat and burning led him to confront the theory of phlogiston. Fire had been domesticated for many thousands of years, but nobody really understood what it was. Heat a piece of wood and suddenly something red and hot begins to flow out of it. The flames cool into smoke and drift away. What remains, the ash, is fragile and much lighter. The name for the substance that emerged as flame was phlogiston. Phlogiston, people concluded—seemingly quite logically—was what gave wood its strength; it could be driven out of wood by heating. All matter, the theory held, contained phlogiston. The theory was convenient enough except for the fact that some materials get heavier when heated. Phosphorus, for example, burns quite readily, but the "ash" that remains weighs more than the original phosphorus. Thus was born the notion that some phlogiston has negative weight. Drive it off by heating, and the material that remains gets heavier.

Negative weight seemed like nonsense to Lavoisier, so he devised another experiment that involved heating substances in closed containers. He would allow nothing to emerge that he could not weigh. This time he heated lead and tin. Both were known to form a crumbly surface layer known as calx. When scraped off and weighed, the calx was heavier than the amount of metal lost from the original sample. The existing idea was that phlogiston with negative weight had been driven out of the metal, leaving behind the ashlike calx. If the metal had gained weight in the form of calx, Lavoisier reasoned, it was not because of losing negative weight but because of gaining positive weight. It had to have come either from the vessel or from the air trapped inside it. When Lavoisier opened the vessel, there was a whoosh as air rushed in. The vessel still weighed the same. Clearly, he concluded, the calx formed as the metal combined with something from the air. Nothing had gone out of the metal.

Through experiments such as this, Lavoisier not only demolished the phlogiston theory but established that combustion resulted from the combination of oxygen with the substance being burned. He even experimented with animals, putting them in closed vessels containing air, pure oxygen, or pure nitrogen, and measuring their heat output. He concluded that, like a fire, living organisms produce heat by using oxygen. Indeed, animals do metabolize food by oxidizing it in a manner that is rather like controlled burning.

As is often the case with scientists who debunk erroneous notions of their predecessors, Lavoisier propounded a few of his own. One was the idea that heat, as well as light, was an element, a basic form of matter. Lavoisier's authoritative and influential textbook called it "caloric," and since he could not explain how matter could behave the way heat sometimes does, Lavoisier said it was an "imponderable fluid." Lavoisier's scientific reputation was so great that his support for the idea of caloric ensured its survival as scientific orthodoxy for half a century more.

Nor was the chemist's reputation forgotten by the leaders of the new French revolutionary government. Two years after his headless corpse was dumped in a mass grave, a regretful government erected a statue of him in Paris.

See PART 1, WHAT IS SCIENCE?; HEAT.

LEAKEY, RICHARD E., 1944– , Kenyan paleoanthropologist. See HUMAN EVOLUTION.

LEUCIPPUS, 5th century B.C., Greek philosopher.
See ATOM THEORY; DEMOCRITUS.

LEEUWENHOEK, ANTON VAN, 1632–1723, Dutch
microscopist—Leeuwenhoek did not invent the microscope, but he did improve the method of grinding lenses
so much that he was able to make major increases in magnifying power. Through microscopes with a magnifying
power of up to 270 times, Leeuwenhoek became the first
human being ever to see single-celled organisms such as
protozoa and bacteria. Until Leeuwenhoek, no one knew
that there were plants and animals smaller than the tiniest insects visible to the naked eye. Even insects as big as
fleas were thought to be crude organisms, but the Dutch
microscopist found "this minute and despised creature" to
be "endowed with as great perfection in its kind as any
large animal." He even found that still smaller "beasties"
live upon the bodies of fleas, and that bacteria live upon
those organisms. The discovery inspired Jonathan Swift,
who pronounced flea as "flay," to compose the following
bit of doggerel:

> So, naturalists observe, a flea
> Has smaller fleas that on him prey;
> And these have smaller still to bite 'em,
> And so proceed *ad infinitum.*

Leeuwenhoek was not a full-time scientist. He did not
even have a university education. He was a janitor for the
city government of Delft in the Netherlands. His hobby
was making microscopes and looking at things through
them. Many others in Europe had microscopes, but no one
knew how to make the lenses as powerful as did
Leeuwenhoek. With his microscopes, Leeuwenhoek examined practically everything he could find, from insects to
animal hairs to rain water. He even examined the scum
from between his teeth and discovered such teeming microbial life in it that he wrote "all the people living in our
United Netherlands are not so many as the living animals
that I carry in my own mouth this day."

At first Leeuwenhoek kept his discoveries to himself. He was a suspicious and private man. But eventually he allowed his friends to look through his lenses. Although his neighbors were more likely to ridicule Leeuwenhoek's curious passion for lenses and microbes, he came to the attention of a local scientist with connections to England's Royal Society—the nerve center of European science. Through him the outside world learned of the wonders this obscure Dutch janitor was discovering. Intrigued, the Royal Society offered to publish Leeuwenhoek's findings. Over the next fifty years, until his death at the age of ninety-one, Leeuwenhoek flooded the Royal Society with hundreds of letters. They must have been among the oddest ever published by the august body—rambling, chatty epistles that were as likely to include Leeuwenhoek's opinions of his neighbors as his descriptions of how red blood cells (Leeuwenhoek discovered these, too) flowed in the capillaries of a tadpole's tail. They were all in Dutch, then considered a lower-class language, and not the Latin that educated people used. Science's first knowledge of bacteria, for example, came in the form of a Leeuwenhoek paper, translated by the Royal Society and published by them in 1683.

In the early days, some of Leeuwenhoek's observations met with skepticism. The notion of whole menageries teeming in a drop of water struck some people as bordering on the preposterous. Naturally, a number of people wanted to verify the observations for themselves. So they wrote to Leeuwenhoek, asking about his methods of grinding lenses. How, they wondered, could he make them so much more powerful than anything an Englishman could make? Leeuwenhoek refused to say.

Stymied, the Royal Society turned to two of the leading English microscopists, Robert Hooke and Nehemiah Grew, commissioning them to try to build better microscopes than any Englishman had made before. Hooke made a microscope that confirmed Leeuwenhoek's observations and established the Dutch janitor as a true pioneer.

See GERM THEORY; HOOKE.

LIFE—It is not as easy to define life as it might seem. The best that most biology texts can do is list some of the traits of living things: they are composed of one or more cells containing complex chemicals, they reproduce their own kind, they grow and develop by incorporating molecules taken in as food, and so on. This, however, does not tell us what life itself is, except that it is the intangible attribute common to objects that possess these traits. But even these qualities of life pose intriguing problems of a philosophical nature.

As soon as life is gone from a living thing, for example, the exercise of the traits ceases irrevocably. But there are organisms that can cease exhibiting most of these traits for decades and yet not be dead. And there are objects, such as viruses—which many people regard as alive—that meet none of these criteria.

As examples of the former, consider a group of small, multicelled animals (including the rotifers, tardigrades, nematodes, and some others) that cope with adverse environmental conditions by going into suspended animation. These small animals, which are common in mosses and soil everywhere, contend with a drought, for example, by simply drying up and resting utterly immobile for years, and even for decades. They stop reproducing, they stop growing, they stop eating, and they stop all of their internal chemical reactions. And these functions stay stopped until moisture returns. Then these animals soak up a little water and stroll off looking for food. The animals must be considered alive during their long dry stage if we are to avoid saying that life has spontaneously arisen in them when they resume activity. But if this is the case, life consists only of a certain arrangement of molecules.

The problem of whether viruses are living is related to this. A virus by itself consists of a small quantity of DNA (or RNA) wrapped in a protein coat. In this form, however, a virus can do nothing. It cannot grow or metabolize and it cannot reproduce. Thus, viruses, by all the usual definitions, are not alive; they are merely very complex chemical compounds. All they can do is ride passively as other forces move them about. If a virus gets near a certain kind of living cell, however, the proteins in its coat will open a hole in the cell membrane and allow the virus's DNA to enter the cell. The DNA then commandeers the metabolic machinery of the cell, causing many new viruses to be manufactured. While they are in a cell, the viral genes behave like those of a living organism, but their goal is simply to produce more viruses that escape the cell to resume the passive existence typical of viruses until they encounter another cell.

See DNA, HOW IT WORKS; CELL THEORY.

LIFE, ORIGIN OF—This is probably the second most challenging mystery facing science. (The origin of the universe is the other.) In a nutshell, nobody knows how life began and it is no help to say, as has become popular in some circles, that life was seeded on Earth by living creatures from elsewhere. That merely changes the location of the mystery. In any case, hardly any biologists are persuaded by this extraterrestrial hypothesis.

Efforts to find out how life began on Earth start with a few well accepted facts. First, the theory of evolution shows how all the present forms of life could have arisen from a much simpler, one-celled ancestor. For another thing, that ancestral cell must have appeared more than 2 billion years ago. There are obvious fossils that look like one-celled algae in rocks of that age. Life may even date back more than 3.5 billion years. Certain rocks of that age from Greenland and southern Africa contain objects that seem to resemble primitive bacteria or algae. Many biologists accept them as such, which means that life probably began less than a billion years after the Earth itself formed, or not long after the molten planet had cooled enough to allow water to remain in the liquid state. This early appearance of life suggests that it was not the astronomically improbable event that religious creationists claim. Life seems to have arisen about as soon as the

Earth cooled to a tolerable temperature. In a sense, life was inevitable.

Another fact that illuminates the mystery of how life began is that many of the complex molecules that are part of the chemistry of life can arise through entirely nonliving processes. In other words, many of the components needed to make a cell could have arisen before life began. Ordinary chemical reactions will produce dozens of different organic compounds—substances made up of carbon and other atoms, and which are found in living organisms or related to substances found in living matter. In fact, at least thirteen organic compounds have been detected drifting in outer space, apparently formed there by spontaneous chemical reactions. Many more can be formed in the laboratory through completely nonbiological processes resembling those that almost certainly existed on the primordial Earth. This was dramatically illustrated in a now-classic 1953 experiment by Stanley Miller and Harold Urey, both American biochemists. They took three gases that were common to the ancient atmosphere—hydrogen, ammonia, and methane—sealed them in a flask with water at the bottom, and zapped electric sparks through the gas vapors. The zapped vapors were then bubbled through the water. The idea was to simulate lightning in the primordial atmosphere, and absorption of the resulting material into the water. After a week the water had turned brown. When they analyzed the water, Miller and Urey found that it contained a mixture of several different amino acids, some fatty acids, urea, and some sugars. Amino acids are relatively complex compounds that, when chained together, form protein molecules. Much the same organic soup can be produced by heating Miller and Urey's original gas vapor mixture, or simply by exposing it to ultraviolet light, a component of sunlight. Similar experiments done more recently by Cyril Ponnamperama, a Sri-Lankan-American biochemist, have shown that all the components of DNA—called nucleotides—also form spontaneously.

In the 1960's Sidney Fox, another American biochemist,

took the story several steps further. First, he found that gently heating a mixture of dry amino acids causes them to link up, forming long chains that would be called proteins if a living cell had made them. Fox called them proteinoids. Second, he found that if the proteinoids are put in water, they assemble themselves into spheres the size of bacteria. He even found that if he let the spheres stay in the water with more dissolved proteinoids, they would undergo a kind of reproduction. Small buds appeared on the surface of the spheres, broke away, and grew. Curiously, the spheres always stopped growing at a certain size, and then grew buds. All of the steps Fox used to make his proteinoids and microspheres could have happened on the primordial Earth. Lightning or sunlight could have made the amino acids. Water containing the amino acids could have washed up on a volcano's warm flanks, causing the amino acids to link up into proteins. Rain or another wave could have wetted the dried protein residue on the volcanic soil, causing it to form spheres.

Although Fox's proteinoid microspheres are not really living cells, they show, along with the Miller-Urey experiment, that many of the attributes of living organisms are inherent in nonliving molecules and their reactions. The molecules that form under ordinary nonliving conditions do not need genes to tell them how to organize themselves into cell-like structures. They are self-ordering.

Before life as we know it can truly exist, however, it must have genes, the master molecules that direct the functioning of the cell and are capable of replicating themselves, so that the control functions that run the cell, and the plans for making the substances needed in order to survive, can be passed on to the cell's offspring. Today's cells rely on DNA for this function. And while molecules resembling some of the constituents of DNA have been produced without life, and whole DNA has been synthesized in the laboratory using bottled ingredients rather like those that can form without life, this synthesis has so far always required enzymes derived from living cells.

Although none of these experiments proves anything

about how life really arose, they do show that a great many of the steps toward life are the result of quite ordinary chemical reactions—reactions that work every time and undoubtedly did work many times in the past. Whether they are the reactions that helped give rise to life we will probably never know. That they *could* have is inescapable.

Advocates of a supernatural origin of life like to point to the complexity of a modern cell and to calculate the odds that a bunch of molecules bouncing randomly around would happen to come together in precisely the right combination. They are quite correct in saying it would take billions of times longer than the universe has been in existence. Where they go wrong is in assuming that the entire job must be done randomly. The Miller-Urey experiment shows that amino acids don't form randomly. Every time their experiment is done it always yields amino acids, and billions of them. The same is true of Fox's experiments with proteinoids and microspheres, and with the other experiments on DNA components. What the oddsmakers also overlook is the fact that the first living cell was probably a very primitive organism. After all, without competitors or enemies it didn't have to be very good at anything. It probably was nothing more than a few specialized molecules inside a container. But as long as it could reproduce, it served as a foundation for the next steps building the complexity of life; they could simply be added one at a time through randomly occurring processes. According to the creationists, the gain of each additional improvement would require that all of the existing attributes of the primordial cell would also have to be recreated through chance events all over again. That isn't how evolution works. It keeps the winning combinations from previous throws of the dice, and adds to them only when subsequent throws of the dice turn up a winner.

See EVOLUTION; CELL THEORY; DNA, HOW IT WORKS.

LIGHT—Though light floods the world around us during most of our waking hours, its nature remains one of science's profoundest mysteries. Light poses a fundamental paradox because tests to determine what it is have led to two contradictory interpretations—that it consists of particles and that it is a wave—neither of which can be shown to be false.

One set of tests shows light to be made of particles called photons, which flow out of the sun or a light bulb. The particles are absorbed by some surfaces and reflected by others. Those particles that enter our eyes hit nerve endings and trigger electrical signals that go to the brain. The particle theory is supported by experimental findings that show the energy of light to exist in discrete particles, or units, called quanta.

Another set of tests shows light not to be particles at all, but waves or vibrations in some medium, just as water waves are vibrations in the surface of a body of water. The wave theory seems compelling, because experiments with light beams show that they behave exactly like water waves or sound waves in their ability to interfere with one another. Take sound waves, for example. Two pure tones played through a loudspeaker can be adjusted so that if they are in phase, they reinforce one another, producing a louder sound. On the other hand, if the waves carrying the tones are exactly out of phase—one wave rising as the other is falling—they will cancel each other out, producing silence. This is exactly how light behaves.

The problem with the wave explanation for light is that waves need some medium through which to propagate. Sound, for example, is a vibration in air. If light is a wave, it must be a vibration in some material substance. For many years scientists assumed that such a substance had to exist, and called it ether. But a classic experiment in the late nineteenth century proved that there was no ether. That result led to the acceptance of Einstein's relativity theory, but left light a puzzling phenomenon with a dual nature, both wave and particle.

Since visible light is only a small part of a much larger spectrum of related waves or particles, the paradox applies to the entire spectrum, which physicists call the electromagnetic spectrum. It includes heat waves, radio waves, x-rays, and related phenomena.

See Entry 12 in Part 2; ELECTROMAGNETISM; QUANTUM THEORY; NEWTON (who suggested light was made of particles); MICHELSON-MORLEY EXPERIMENT; RELATIVITY (in which the speed of light is the only absolute); EINSTEIN.

LIGHTNING—See ELECTRICITY.

LIGHT-YEAR—A unit of distance, not time. It is the distance a beam of light will travel in one year. Because light moves at 186,000 miles per second, a light-year is roughly six trillion miles.

LINNAEUS, CAROLUS, (Carl Linné), 1707–1778, Swedish botanist—Linnaeus created the system of scientific names that biologists all over the world use today to identify living species. The system calls for a genus name first and a species name second. *Homo sapiens*, for example. The system is called binomial nomenclature. It may not seem like much of an achievement, but it came as an enormous boon to biologists of the eighteenth century, who in some cases were struggling with dozens of different local names for the same kind of plant or animal.

Such confusion, which could easily have caused crucial damage in the study of disease-causing organisms, might have continued to this day had not someone come up with a system that everyone found acceptable.

Under the Linnaean system, the first name of an organism defines the genus to which it belongs, which may include several kinds of closely related organisms, while the second name defines its species, which ordinarily includes only that one particular kind of organism. Thus, within the genus *Homo* there are at least three known species, *Homo habilis* and *Homo erectus* (both of which are extinct), and *Homo sapiens* (which is not yet extinct). However, no system is perfect, and it has happened that after names have been assigned, subtle distinctions have been found within a species. As a result, the species has been divided into subspecies. When this happens, the subspecies that is represented by the "type specimen" (the specimen, in a museum or elsewhere, that is declared to be the official exemplar of the species) is given a subspecies name identical to the species name. For example, living human beings are called *Homo sapiens sapiens*, and our close relatives colloquially known as Neandertals are officially called *Homo sapiens neanderthalensis*.

One of the traditions established by Linnaeus was the use of Latin or Latin-like names for classifying organisms. At one time Latin was the language of scholars throughout much of Europe. It neatly avoided nationalistic sensitivities that might attach to the use of a "living" tongue, and provided a neutral, common means for scholarly communication. Important scientific works were written in Latin. Some enthusiasts of Latin even went so far as to latinize their own names. Thus Carl von Linné called himself Carolus Linnaeus, and the Pole Kopernik became Copernicus. With the decline of classical scholarship, English has replaced Latin as the global language of science, and new species are given all sorts of names, as often Greek as Latin (making purists shudder), or only thinly latinized names. Still, the system of binomial nomenclature continues to serve biology well.

In setting up his biological classification system, which included the more inclusive categories of family, order, phylum, and kingdom, Linnaeus took for himself the privilege of selecting the names for the species that he knew. It was he who chose the Latin *homo*, for man, presumably unaware that it might someday be confused with the Greek *homo* which means "same." Linnaeus did correctly classify the whales as mammals, and he recognized that human beings were closely related to apes. But he went too far by present standards when he named the orangutan *Homo troglodytes*. The name has since been changed to *Pongo pygmaeus*.

See TAXONOMY.

LUCRETIUS, 96?–55 B.C., Roman poet. See ATOM THE-
ORY.

LYELL, CHARLES, 1797–1875, British geologist—
When Charles Darwin set sail on his epochal exploration
aboard the *Beagle*, he left England a believer in the biblical view of creation, but he also left with a new copy of
Charles Lyell's book, *Principles of Geology*. Through the
book, Lyell taught Darwin to look at the world around
him with new eyes. As a result Darwin saw the evidence
that would lead to his theory of evolution. Chief among
Lyell's influential ideas was the view that the Earth is
very old—vastly older than most people supposed—and
that the processes that formed its geological features had
been operating very slowly for a very long time. In fact,
Lyell said, the very same processes are still operating at
their usual rates, a concept called uniformitarianism. It's
just that the rates are so slow that the processes are virtually imperceptible.

Until Lyell published his book, most thinking people accepted the idea that the Earth was young, and that even
its most spectacular features such as mountains and valleys, islands and continents were the products of sudden,
cataclysmic events, which included supernatural acts of
God. In its scientific guise, this theory was known as catastrophism, and one of the archetypal catastrophes was
Noah's flood.

While Lyell's theory was quite simple, he backed it up
with massive documentation gathered from a lifetime of
field observations. Lyell had made geology expeditions all
over the British Isles, Europe, and eastern North America. For example, he analyzed the layers of rock in mountains, including the cracks and folds in these layers, and
reasoned that the only way they could have formed was
through various gradual processes or frequently repeated
events, such as earthquakes and volcanic eruptions. In
certain high mountains, for example, Lyell found the same
type of rock that results when sediments compress at the
bottom of a lake or ocean. Some mountaintop rocks even

had seashells embedded in them. They could not have reached their lofty positions except by having been pushed up from below. Volcanic mountains, by contrast, showed signs of having been built up by the addition of successive layers on top.

Lyell also recognized that different layers of rock contained different complements of fossils. He reasoned that as the geologic changes had progressed, they had changed the local environment and forced the extinction of existing species while making way for new kinds of plants and animals. Lyell did not attempt to explain how the new species had emerged. Later, Darwin took care of that detail.

Lyell came to his work in geology by a back route. His father, a wealthy landowner and lawyer, wanted the boy to be a lawyer too. Lyell did study law, but long periods of close reading gave him pain in his eyes, and several times he took leave of law school to wander about the countryside, indulging a personal passion that, in his day, was known as geologizing. Basically it meant rambling about the hills and valleys, examining their structure and enjoying the outdoors. Lyell's father repeatedly prodded him back into law school, and he eventually got his degree and was admitted to the bar. But his passion for geology grew and he devoted more and more time to it, eventually producing the famous three-volume work that Darwin took on the *Beagle*. Actually, the basic idea of uniformitarianism was not original with Lyell. It had been put forth by James Hutton, a British geologist, who died the year Lyell was born. Hutton's ideas never caught on, though, until Lyell reintroduced them with massive documentation from his field studies. Later, Darwin used the same style of massive documentation in his own books, thus winning support for a theory that others had suggested but not demonstrated.

In attempting to find order in the many jumbled layers of rock he observed, Lyell examined the fossils contained within them and found characteristic combinations of species that seemed to represent distinct eras in geologic history. He gave the different eras the names that geologists and evolutionists recognize today—Eocene, Miocene, Pliocene, and Pleistocene.

Lyell worked most of his scientific life as an independent scholar and not as a professor. He devoted most of his time to revising his three-volume masterpiece through successive editions, each time adding new data and answering critics. Curiously, although he provided the intellectual framework for Darwin's ideas, Lyell did not accept the idea of evolution until after he read *On the Origin of Species*. Thereupon, Lyell added evolution to the next revision of his *Principles of Geology*, and even added some new arguments of his own, thus helping to win more support for Darwin. Lyell was working on the twelfth edition of the book when he died in London in 1875. He was buried in Westminster Abbey.

See UNIFORMITARIANISM; EVOLUTION; DARWIN.

MACROMOLECULE—This term means "big molecule." It is a term of convenience used in discussing big molecules that are made up of smaller ones. Macromolecules may consist of anywhere from hundreds to hundreds of thousands of smaller molecules. Examples include DNA, plastics, and proteins.

MAGIC BULLET—A famous colloquial term for a drug that would go straight to its target, the diseased tissue, and confine its effect on that tissue alone. Real drugs, by contrast, usually affect the whole body, producing unwanted side effects. The term "magic bullet" was coined by Paul Ehrlich, the German physician. Lately, researchers have raised the prospect of attaching drugs to tailor-made antibodies, with the hope that the antibodies, which bind only to specific tissues, will concentrate the drug at the desired site of action.

See EHRLICH; MONOCLONAL ANTIBODY.

MAGNETIC MONOPOLE—A hypothetical particle of subatomic proportions that carries the force of magnetism. A monopole would be an isolated north or south pole. No one has ever detected a monopole, but one theory suggests monopoles should exist if magnetism is like electricity, which is carried by particles that have either negative or positive charges. The concept of a monopole is based on the fact that magnetism and electricity are two manifestations of the same fundamental phenomenon, electromagnetism. Many physicists, however, doubt that monopoles exist.

See ELECTROMAGNETISM; FARADAY; MAXWELL (the physicists who linked electricity and magnetism).

MAGNETISM—Few everyday phenomena are more perplexing or more fascinating than the forces of magnetic attraction or repulsion. There is something deeply startling for even the most sophisticated mind as it contemplates the ability of a chunk of iron to suddenly lift off the ground and fly into the air, attracted to a magnet, or that one magnet can push another away, even though both phenomena are usually part of child's play. How does one lifeless piece of metal "know" anything about the presence of another lifeless piece of metal?

The fact is that magnets do indeed send something invisible into the space around them, something that some other metals can "feel." It may be subatomic particles flowing out of one pole of a magnet, looping around through space, and reentering the magnet at the opposite pole. The existence of these particles, dubbed magnetic monopoles, has been suggested but never demonstrated. What is clear, however, is that some effect does travel through space from one pole to another. Physicists call this effect a field. Its strength and direction can be measured at any point within it.

Iron, cobalt, and nickel can all be made into magnets. This is because their atoms, each of which is a tiny magnet, can be aligned in the same direction, so that their north poles all point in the same direction. Normally, the orientation of the atoms is random, and the magnetic forces of the various atom-sized magnets cancel each other out. Once the atoms are aligned, however, the combined strengths of their fields produce the familiar effects of the toy magnet.

See ELECTROMAGNETISM (the basic force of which magnetism and electricity are just different manifestations); MAGNETIC MONOPOLES.

MALPIGHI, MARCELLO, 1628–1694, Italian histologist—Malpighi was to the microscope what Galileo was to the telescope. He didn't invent the instrument, but he used it to make far more detailed and sophisticated investigations than anyone had before. And in the process he opened whole new fields of study for science.

Malpighi used the microscope to study smaller structures and processes within the body—both human and animal—than anyone had ever seen before. He discovered the capillaries, the tiny blood vessels that carry the blood from the outermost reaches of the arteries to the veins for the return trip to the heart. In so doing, he vindicated William Harvey's assertion that blood circulates in the body. Several microscopic structures of animal cells are named for Malpighi. These include certain structures within the kidney that filter blood, and the layer of skin in which new skin cells are generated. Malpighi also studied the development of the heart in chick embryos, the respiratory systems of insects, the cellular structure of leaves, and a host of other microscopic domains within living organisms. In fact he opened up so many new avenues of investigation that he is sometimes considered the father of histology, the study of cells.

Like Galileo's work, Malpighi's ran afoul of the ancient traditions of medieval philosophy, but in so doing helped launch the coming world of experimentally founded science. Many of his findings encountered hostile receptions, and he twice had to leave universities where he taught when he found the opposition too disturbing. Unlike Galileo's difficulty, the hostility came not from the church but from professors of medicine who either doubted the truth of Malpighi's minute discoveries or who ridiculed their relevance to medicine. Malpighi was reduced to asserting that although he might not know how the tiny structures related to disease and health, he was sure they did. It was a position not unlike that of scientists today who say that although basic research may yield no immediately obvious practical benefits, they are sure that it ultimately will.

Malpighi died in Rome of apoplexy.

See HARVEY.

MASS—This is the property of an object that gives it weight when gravity pulls on the object. On the surface of the Earth, an object's mass is defined as being equal to its weight. If the same object were on the moon, however, its mass would be the same as on Earth (there would be no loss of matter), but its weight would be only one-sixth as much. The reason why objects weigh less on the moon is that weight is the result of the mass of both an object and of the larger body that is pulling on it through gravity, and this gravitational pull is greater for a greater mass. Therefore, the Earth, having much more mass than the moon, exerts a stronger pull on objects. It is true that a ball tossed into the air exerts a gravitational pull on the Earth, and that the Earth, in turn, is pulled an infinitesimally small distance toward the ball. However, the Earth's effect on the ball is overwhelmingly greater.

The mass that objects have is the sum of the masses of all their atoms. Nearly all of an atom's mass is concentrated in the protons and neutrons of its nucleus. Thus, when you weigh yourself, more than 99.99 percent of the weight you see on the scale is due to the weight of just two things: protons and neutrons.

See MATTER; WEIGHTLESSNESS; ATOMIC STRUCTURE; RELATIVITY (which shows that an object's mass can increase as it approaches speeds close to that of light).

MATTER—One manifestation of a mysterious phenomenon that may also show itself as energy. Even in the familiar, hard form of molecules and atoms, matter actually consists of tiny bundles or packets of energy called quarks, which make up the protons and neutrons, and electrons. However, most of what seems to us to be solid material is actually empty space. The electrons are actually farther apart, in scale, within a single atom than the planets of our solar system. If it were not for the forces binding these far-flung particles, one atom could pass through another with the chance of a collision between the two being quite remote; in fact, it would be no trick to walk through a wall. The only reason we can't do this is that the forces binding the wall's subatomic particles (which binds them into atoms and the atoms into molecules) repel the forces binding our bodies' particles together.

Antimatter is identical to matter except that its particles have opposite electrical charges. It is only by convention that we speak of our own world as one of matter. It could just as well be called antimatter, and we could speculate about the existence of other worlds made entirely of matter. When particles of matter and antimatter meet, they annihilate one another, releasing the energy packets of which they are made.

See ATOM; ATOMIC STRUCTURE; MOLECULE; PARTICLE; RELATIVITY (wherein Einstein showed matter and energy to be alternative forms); QUARK; ANTIMATTER.

MAXWELL, JAMES CLERK, 1831–1879, Scottish physicist—If any major figure in the history of science has been underrated by the general public, it is James Clerk Maxwell. Many physicists would rank him close to Isaac Newton, but most nonscientists have probably never heard of him.

Maxwell is chiefly remembered for his work in two areas—establishing the relationship between electricity and magnetism (a relationship that underlies all modern applications of electricity and electronics), and showing that heat was not a mysterious fluid but a property of molecules in motion. Maxwell's work was chiefly mathematical, drawing on other people's experiments. There is a fundamental acceptance in science that if observed phenomena can be expressed in mathematical equations, they must be the result of orderly processes in nature. The equations can be tested by changing the value entered for one factor, doing the necessary calculations to solve for the other factors in the equation, and then conducting an experiment that parallels the equation. If, for example, doubling the voltage in the equation yields a certain change in some other factor, the experiment would involve doubling the voltage in a real circuit and seeing whether the real changes are as the equation predicted. Maxwell's mathematics are beyond the scope of this book, but an appreciation of his work is not.

Maxwell's work on electromagnetism, a term he coined, followed the experiments of Michael Faraday, a largely self-taught British physicist who was a brilliant experimentalist but lacked mathematical ability. Faraday had shown that moving a magnet near a wire caused electricity to flow in the wire (the basis of electrical generators), and conversely, that electricity flowing in a wire would cause the wire to rotate about a magnet (the basis of the electric motor). Faraday surmised that electricity and magnetism were two manifestations of the same fundamental natural force.

Maxwell's mathematical work not only confirmed the relationship between electricity and magnetism, but revealed that electromagnetism was propagated through space as waves. Moreover, he calculated the speed with which the waves moved, and found it to be very close to the speed calculated for light. Maxwell correctly surmised that light was a kind of electromagnetic wave, along with such invisible forms of energy as infrared and ultraviolet radiation. He concluded that the only differences between these physical phenomena were differences in wavelength and frequency. In other words, some electromagnetic waves are more tightly packed and pass a stationary point as rapid vibrations, while others are more drawn out and pass as slow undulations.

We know today that devices capable of detecting differences in electromagnetic waves—whether eyes responding to light or tuners responding to a radio station—are simply devices that are sensitive only to certain wavelengths or frequencies, and which then produce a secondary effect of this sensitivity—such as color perception in the case of eyes or sound in the case of radios.

Maxwell's other major contribution to science was what is known as the kinetic theory of gases. Until Maxwell's time many people had thought that heat, for example, was an invisible fluid, called caloric, that somehow got into objects. It was a reasonable guess, since heat does flow from hotter objects to adjacent, cooler ones until both reach the same temperature. An alternative theory held that heat was only a property of gas molecules in motion. Maxwell helped abolish the caloric theory with a series of equations that turned out to describe the physical phenomenon of heat in a gas more aptly. Maxwell said that the temperature of a gas was a statistical average of the speeds at which individual gas molecules moved about. He maintained that when a volume of gas is heated, the heat energy is absorbed and used to increase the motion of the gas molecules. The gas molecules literally whiz about faster and bump into one another harder. They also bump into the walls of the container harder and more frequently, increasing the pressure exerted by the gas. Maxwell's insight was to show that the molecules need not all move at the same speed. A thermometer simply averages out the energies of individual molecules. The word kinetic, as in kinetic energy or Maxwell's kinetic theory, refers simply to motion.

To help explain his theory in nonmathematical terms, Maxwell invented one of the more famous conceptual devices in science—Maxwell's demon. Imagine two boxes containing the same gas at the same temperature. Between the boxes is a small door controlled by a little man, or a demon, as Maxwell put it. The demon watches the molecules moving about in the first box, and every time he sees a fast one coming toward the door, he opens it and lets the fast molecule enter the second box. If the demon sees a slow molecule coming from the second box, he lets it into the first box. As more fast molecules move into the second box, the temperature in the box will rise. As slow molecules accumulate in the first box, its temperature will fall. Maxwell's demon obviously does not exist, but if it did you could heat or cool a room without using any additional energy.

Maxwell was born in Edinburgh in 1831 and grew up at his family's country house. Though fairly well off, young Maxwell was very much a country boy, and when he went off to school in the city, his country clothes and ways marked him as something of a bumpkin. He was severely taunted by his more urbane classmates, and developed a very shy and introspective nature. Still, by the age of nineteen he had had two scientific papers published by the Royal Society. He held several teaching posts at colleges, eventually reaching King's College in London, where he did most of his best work within a five-year span. Country life beckoned him when he was thirty-three and he retired to his family's country estate in Scotland. He lectured occasionally at Cambridge, and at the age of forty was persuaded to return to academic life and to establish the Cavendish Laboratory at that university. It quickly became one of science's leading research centers and remains so today. Maxwell, however, fell ill at the age of forty-eight and died.

See ELECTROMAGNETISM; WAVES; FARADAY.

MAXWELL'S DEMON—James Clerk Maxwell invented this imaginary character to help explain one of his insights into the nature of heat in gases. The demon is a little man who controls a door between two chambers of gas that have the same temperature. He allows fast-moving gas molecules to go one way through the door and slow-moving molecules to go the other. Eventually one chamber accumulates more fast-moving molecules and becomes hotter, while the other chamber gets more slow molecules and cools down. The demon was Maxwell's way of explaining that temperature is a measure of the average energy possessed by individual molecules moving at different speeds. A fuller explanation is in the Maxwell biography, above.

MECHANICS—This is a broad, catchall term for several classes of phenomena involving objects in motion. The term is sometimes used to refer to the force-multiplying effects of simple machines, such as levers and pulleys. But in a larger sense, it can also refer to Newtonian mechanics, the laws of motion that Isaac Newton derived to show that the force holding the planets in orbit is the same as the force that causes an apple to fall. Einstein showed that Newton's laws are only approximations. For most commonly observed phenomena, Newtonian mechanics gives an answer that is indistinguishable from that of Einsteinian mechanics, which is embodied in the theories of relativity. However, when objects are moving at an appreciable fraction of the speed of light, the error in Newtonian mechanics becomes large. Relativity theory gives the right answers for objects moving at all speeds.

See NEWTON; EINSTEIN; RELATIVITY; ARCHIMEDES (who explored the mechanical effects of simple machines).

MEIOSIS—This is the process that special cells in the gonads undergo to produce sperm or ova. It is distinguished from mitosis, in which cells divide and produce two identical offspring, each with a full set of paired chromosomes. Mitosis is responsible for the ordinary process of development in which a fertilized ovum grows into a fully developed individual with billions of cells. In meiosis the cells divide, but each daughter cell, whether a sperm or ovum, gets only one chromosome from each pair of chromosomes in the parental germ cell. At conception, two half-sets of chromosomes—or one chromosome from each parent—combine to produce a full set that will begin developing.

See MITOSIS.

MEMBRANE—Life is a special state of matter in which the levels and compartments of organization are vastly more complex than in nonliving matter. A star or planet can have a considerably complex internal structure, but even a worm has many times more internal complexity. Membranes are the reason why life can be so complex.

The most common membrane is the bag that encloses every living cell. This cell membrane, like all membranes (including artificial ones made for industrial uses), is made so that certain kinds of molecules can pass through it while others cannot. As a result, cells can easily take in water and various nutrients (which are usually relatively small molecules) but not lose the larger protein molecules that they make for their own specialized purposes.

Cells modify the properties of their membranes so that the latter are, in a sense, intelligent gatekeepers. The membranes of cells in the adrenal gland, for example, let out molecules of the hormone adrenaline—a protein with properties the membrane recognizes—but keep in other proteins needed to help synthesize adrenaline. Most cell membranes are studded with receptor molecules shaped precisely to fit some outside molecule. When this molecule comes along, it will nestle into the receptor, which then will help pull it through the membrane into the cell. In some cases, very small molecules go in and out of cells simply because they can fit through pores in the membrane.

Groups of cells (tissues) may also produce a membrane that surrounds the group and performs similar functions to those of the cell membrane.

See OSMOSIS (one mechanism that regulates passage of molecules through a membrane); RECEPTORS.

MENDEL, GREGOR, 1822–1884, Austrian geneticist— People have always known that children resemble their parents, but until the twentieth century, most attributed this fact to a blending of bloods from the mother and father. So common, in fact, was the blood theory of inheritance that our language still uses the phrase "it's in the blood" to explain that a trait is inborn. If Dad was tall and Mom was short, Junior usually turned out medium. It seemed obvious that this was the result of a blending of the two parental bloods.

Not until 1865 was there evidence to the contrary, from Gregor Mendel, an Austrian monk, and even then the evidence lay almost unnoticed until 1900 when, in one of the great coincidences of science, it was rediscovered by three scientists, each working independently. Mendel's experiments with pea plants proved that heredity was transmitted not by anything that could blend (like blood), but by something that passed from parent to offspring in discrete and unchanging particles. Once rediscovered and confirmed, Mendel's findings set off a search for the exact nature of the particle. The search first zeroed in on the cell nucleus, then on the chromosomes within the nucleus, and finally on the molecule called DNA, of which the chromosomes are chiefly made. In the 1950s biologists at last learned the precise molecular structure of the particle that had for decades been called the "gene."

Mendel, an Austrian monk, did his experiments in the monastery garden at Brünn, which later became part of Czechoslovakia and was known as Brno. He worked with several varieties of garden pea plants, varieties that showed two alternative forms of various features. For example, one variety was tall and another short. Some had white flowers while others had purple. There were varieties that produced smooth seeds and others that produced wrinkled seeds. Mendel concentrated on seven sets of such traits, each of which came in two forms. He found that by crossing one variety of pea plant with another (pollinating one variety with the pollen from a different variety), he could sometimes transfer one or more of these seven traits. But not always. Sometimes the offspring looked exactly like one parent, sometimes exactly like the other. In no case, however, did any of the traits blend.

Mendel became intrigued by this observation, and kept very careful records on his cross-pollinations and their outcomes. For example, he found that every time he crossed a tall and a short pea plant, the progeny were always tall. But when he crossed one progeny plant with

another, he found that only three out of every four offspring were tall. One out of four was short. It resembled
neither parent, but instead looked like one of its grand-
parents.

In the data from hundreds of such experiments, Mendel
perceived a pattern. He created a hypothesis to explain
the pattern, saying that his results could be explained
only if three statements were true: First, that each parent
plant carried two units (we now call them genes) of heredity governing each trait. Second, that when the parent
plant produced its reproductive cells (ovum for the female
and pollen for the male), only one of the two units went
into each cell. And third, that one of the two alternative
genes that governed a trait dominated the other such
gene.

In the case of pea-plant stature, for example, Mendel
found that when a "tall" gene combined with a "short"
gene, the resulting plant was always tall. In other words,
tallness as a trait dominates shortness. But when this new
plant reproduces, it will give half its reproductive cells
"tall" genes and the other half "short" genes. If such a
plant is then crossed with another like it, there will result
only three possible combinations of genes. The following
chart illustrates the situation:

	tall	short
tall	tall-tall	tall-short
short	tall-short	short-short

Three out of four—or three-fourths of the offspring—
will be tall. One because it has only "tall" genes and two
because they have one "tall" gene, which is enough, since
"tall" is dominant. One-fourth of the offspring will be
short because they have only "short" genes.

The same statistical distribution applied for all of the
other traits Mendel studied in peas. Mendel's genius lay in

concentrating on simple traits with only two alternative forms, and in perceiving the statistical relationships between parental traits and those of the offspring. But genius was not Mendel's only useful attribute. There appears to have been a good dose of chutzpah too. A close study of Mendel's reported data shows that his numbers are too good to be true. His counts of the number of plants showing a trait come much closer to the statistical ideal than would be likely if he had been accurately recording his findings. This is akin—had he been calculating the odds on a coin toss—to his having flipped the coin ten times, coming up with six heads and four tails, and claiming to have gotten five heads and five tails, which statistically is the most probable result since coins have only two sides. Mendel apparently "cleaned up" his data to make the numbers fit his theory more closely than they did in reality. Mendel's theory was right, however, and he should not have worried that an imperfect statistical distribution would invalidate it, any more than the coin toss results of six and four invalidate the theory that coins have only two sides.

As it happens, not all traits are governed by the simple two-gene dominance model that Mendel hypothesized. In fact, most are not. Many traits are governed by four, six, eight, or more genes, some of which may have only partial dominance. The result can be the appearance of a "blending" inheritance. Among humans, skin color is a good example of this. So is height. One human example of simple mendelian inheritance, as geneticists call the pattern of heredity seen with the simple two-gene model, is albinism, the inability of one's skin and hair cells to produce melanin, the skin pigment. Albinism is a recessive trait, requiring the inheritance of one recessive gene from each parent, both of whom, though of normal color, must be carrying one recessive gene paired with a dominant gene that is sufficient to carry out melanin synthesis.

See MODERN SYNTHESIS (which linked Mendelism with Darwinism to produce the modern theory of evolution); DARWIN; GENES, HOW THEY WORK.

MENDELEYEV, DIMITRY IVANOVICH, 1834–1907, Russian chemist—Mendeleyev's chief contribution was in devising a system by which the sixty-three chemical elements known in his time could be classified sensibly. He called his system a periodic table of the elements. It turned out to be not simply a classification system, but a guide to the discovery of previously unknown elements and to the very nature of matter itself. The table strongly implied that all of the atoms that comprised the basic units of matter were related, with all of them built according to the same rules.

Mendeleyev was the seventeenth and last child of a small-town Siberian schoolteacher who became blind. To support the family, Mendeleyev's mother ran a glass factory. When Mendeleyev was thirteen his father died and his mother moved the family to a larger town where her son could get a better education in the sciences for which he had already shown a great aptitude. Mendeleyev eventually became a professor at St. Petersburg University in what is now Leningrad.

As a scientist, Mendeleyev was deeply interested in the properties of elements, the chemically fundamental and uniform substances of which all other matter was composed. Some years earlier, John Dalton, an English chemist, had developed the concept of atomic weight—a number that refers to the relative mass of an atom of any given element. Mendeleyev found that if all the known elements were arranged in sequence according to their atomic weights, certain common chemical properties recurred at periodic intervals along the sequence. For example, every eighth element turned out to be a gas that did not combine with any other element—one of the so-called noble gases.

Recognizing this periodicity, Mendeleyev arranged the sequence of elements in stacked rows so that—to follow the example given above—all of the noble gases fell into one column. As it happened, however, some of the elements seemed to fall into the wrong columns. That is, their chemical properties were not like those of other elements in the column. But Mendeleyev had so much faith in the idea that all the elements should be part of an orderly system that he simply moved the misplaced elements to more appropriate columns, leaving gaps. His faith was vindicated years later, when the missing elements were found and fit perfectly into Mendeleyev's gaps.

Not until the twentieth century did physicists pin down the reasons why the elements fit so well into the periodic table. Simply put, it is because they are all built of the same subunits—neutrons, protons, and electrons. The atom of hydrogen, the lightest element in atomic weight, consists of one proton with one electron orbiting around it. Helium, the next lightest element, has an atom consisting of two protons orbited by two electrons. (Actually electrons do not orbit the nucleus of an atom in the manner of a planet orbiting around the sun. Rather, they fly in many pathways all within one or more shells that surround the nucleus at some distance from it.) Still heavier elements have scores of protons and neutrons in their nuclei and scores of electrons orbiting the nuclei. In all but the simplest atoms, these electrons are in different shells. No more than two electrons can occupy the innermost shell, for example, while eight electrons occupy the next shell, and so on.

Mendeleyev had many other scientific interests besides the elements, and worked on improvements in the Russian chemical and oil industries. He even worried that oil was a finite resource that might be depleted. He visited the United States in 1876 and criticized American oil companies for concentrating on simply increasing the volume of production instead of seeking ways of getting more oil out of a well—an approach that has only recently been taken seriously.

Even before the rise of the Soviet system, Mendeleyev, like Russian scientists today, had problems with the government. In 1890 he endorsed a petition to the czar from students at his university who were complaining of various injustices. The government promptly dismissed him from the university and allowed him no further academic posts.

See PERIODIC TABLE; ELEMENTS; ATOMS; DALTON.

METABOLISM—All of the chemical reactions that go on inside a living cell, either to build new, needed molecules or to break down old, unneeded ones. Strictly speaking, metabolism occurs only within cells, but the term can also apply to what an entire multicellular organism does, which is actually only the aggregate of its cellular metabolic processes.

METEORITES AND METEORS—The Earth is repeatedly visited by extraterrestrials—chunks of rock and metal left over from the formation of the solar system that happen to get pulled in by the Earth's gravity. The naming of these objects can be tricky. As long as they are drifting in space (usually in an elliptical solar orbit that crosses Earth's orbit), they are called meteoroids. When they streak through the atmosphere, heated white hot by friction with the air, they are called meteors. Those that are not completely vaporized during their fiery descent hit the ground (or the water, more often) and are then called meteorites. The largest known meteorite was a seventy-ton monster found in South Africa. There are several circular features on the Earth's surface that are probably craters formed when big meteorites hit. Meteor Crater in Arizona is definitely the result of such an impact. The reason that there aren't more such craters is that various geological processes, such as erosion, keep remodeling the Earth's surface. This doesn't happen on the moon, the surface of which is pockmarked with meteor craters.

Meteoroids seem to have two main sources. One group apparently had its origin at the beginning of the solar system, back when a nebulous, disk-shaped cloud of gas and dust was swirling about a central core. Most of the cloud drew toward the center, where the growing clump of matter became steadily stronger in gravity. When the mass of this matter became great enough, the internal pressure ignited the nuclear fires that turned the mass into a star. At the same time, clumps of matter were forming farther out in the disk-shaped cloud, and the largest of these became planets. Left over were countless smaller chunks—most of them mere grains of sand, some car-sized boulders, and a few as big as mountains. The smaller they were the more readily their orbits could be pulled into ellipses around the sun by the passage of other, larger bodies that happened to sail through the solar system and pass right on out again.

As the Earth moves around the sun, it continually encounters new regions through which these chunks are pursuing their own orbits. Our gravity distorts their orbits, pulling some directly toward us to become meteors.

The second major source of meteors includes the comets. These are huge balls of ice and rocky particles, or "dirty snowballs," as it were. When they approach the sun, some of the ice vaporizes (helping form the tail of the comet as the solar wind—a draft of atoms and subatomic particles spewing out of the sun—sweeps it back), releasing a steady cloud of tiny particles that continue in the same orbit as the comet. Thus, every comet's path is strewn with particles chasing the comet's head. When the Earth crosses the orbit of a comet—even though the comet head may be long gone—it pulls in the nearest particles, which occur in such numbers as to produce a meteor shower. In some of these showers one can see 60 to 100 meteors an hour. Since the Earth crosses this point in its own orbit once every year, meteor showers are predictable annual events.

See SOLAR SYSTEM, ORIGIN OF.

MICHELSON-MORLEY EXPERIMENT—A classic experiment performed in the late nineteenth century that disproved the existence of ether, a mysterious substance formerly thought to permeate the universe, and through which light waves were believed to propagate. The experiment was part of the intellectual groundwork of Einstein's relativity theory.

See RELATIVITY; LIGHT.

MICROSCOPES AND MICROSCOPY—Many advances in science are the result not of bright ideas but better instruments. The development of the microscope by Anton van Leeuwenhoek, for example, led to the discovery of worlds hitherto unseen. Robert Hooke's brilliant use of microscopy added to the wealth of detail that radically altered the human sense of scale.

See LEEUWENHOEK; HOOKE.

MILLER, STANLEY, 1930– , American biochemist. See LIFE, ORIGIN OF.

MITOCHONDRION—Mitochondria are sausage-shaped microscopic structures found inside every living cell except bacteria and the closely related blue-green algae. They convert the chemical energy derived from food into a form of energy that the cell can use to power all of its internal activities. Mitochondria take in sugar and oxygen (which arrive from the digestive system via the bloodstream) and "burn" the sugar using the oxygen, to release the energy stored in the sugar's chemical bonds. The sugar and oxygen are broken down and recombined into the waste products water and carbon dioxide. The energy liberated in this process is stored in another molecule called ATP (adenosine triphosphate), which travels out of the mitochondria to other parts of the cell where energy is needed. Once at the work site, ATP gives up one of its phosphates, releasing the energy in the chemical bond that held this phosphate to the rest of the ATP molecule, and becomes ADP (adenosine diphosphate). ADP then returns to a mitochondrion to pick up new energy in the form of a bond that attaches a new phosphate.

Mitochondria are unusual among the components of cells in that they contain their own DNA. They are, in fact, the only structures outside the nucleus where the cell keeps DNA. In some ways mitochondria are like little one-celled organisms imprisoned within a cell. They "reproduce" by dividing, a process that may endow a cell with thousands, even half a million mitochondria, all inheriting copies of the same DNA. The DNA in animal mitochondria, incidentally, is inherited only from the mother, by way of the mitochondria in her ova. Sperm contribute only nuclear DNA to the cell.

Some evolutionists suspect that mitochondria originally evolved as free-living bacteria (which have no nucleus) that entered into a permanent symbiotic relationship with nucleated cells.

See CELL; CELL THEORY; SYMBIOSIS.

MITOSIS—This is the ordinary process of cell division, in which one cell duplicates its entire set of chromosomes and then splits in two, with each new cell getting one of the two sets of chromosomes. It is a complex process of many stages, and biologists still do not fully understand how all the events in mitosis occur. The process of DNA replication, however, was solved in the famous discovery of the double helix.

See MEIOSIS; CELL; DOUBLE HELIX.

MODEL—If you're trying to study some phenomenon that you can't see very well, one way of approaching the problem is to make a model. A model can be a physical object along the lines of the more colloquial use of the term (model car, model of the solar system, and so forth), or it can be a verbal model—a simile, metaphor, or analogy of what you think some system or situation might be (such as the concept that white blood cells engulf bacteria the way amoebas eat their food). Models may also consist of mathematical formulas, often in a computer program. If your mathematical model is good, it will predict what will happen in a real situation on the basis of numerical information you give the computer.

Every area of science uses models as intellectual devices for making nature easier to understand. A model that reliably predicts the outcomes of real events or that continues to fit new data is essentially a kind of theory, a broad statement of how nature works. Sometimes, however, models are so seductively attractive that they live way beyond the time that empirical findings refute them. Consider, for instance, the model of heat as a mysterious fluid called caloric. Some models capture popular attention and hold it until long after scientists have discarded them. The model of an atom as a little solar system is one such example. This model maintains that electrons orbit the atomic nucleus like planets orbiting the sun. It was, however, discarded half a century ago when physicists realized that electrons move every which way inside shells enclosing the atomic nucleus.

See THEORY; CALORIC; ATOMIC STRUCTURE.

MODERN SYNTHESIS—This is the term biologists use for a stage in the intellectual development of their field, in which two formerly separate specialties were recognized as parts of the same larger phenomenon. It describes what happened over a period of about two decades, from the 1920s through the 1940s, when genetics was seen to provide a molecular basis for understanding how evolution works. The modern synthesis—a term coined by Julian Huxley, the grandson of Darwin's great advocate Thomas Henry Huxley—gave the once diversified field of biology a powerfully unifying concept of how the world of life works.

When Darwin put forth his theory of evolution by natural selection in 1859, the theory lacked two important components: Darwin did not know the mechanism by which plants and animals pass their traits on to descendants, nor did he know how variations might arise in those traits. Darwin had asserted that individuals of a species varied slightly in their inherited traits, and that those possessing the more advantageous traits would thrive at the expense of those lacking these traits. Those individuals that thrived would leave more descendants (inheriting the traits) than would their less fortunate kin. Over many generations, Darwin asserted, an accumulation of advantageous traits would produce a population so different from its ancestors that it would constitute a new species. Natural selection, as Darwin called this idea, was a powerful concept, but without some idea of how organisms got their traits or passed them on it did not immediately win many believers. Darwin's idea of gradual evolution did, however, become widely accepted. The slow and steady accumulation of traits accorded well with what naturalists already knew of living animals. If there were a certain bird species in England, for example, they might know of a closely related variety in France and a more distantly related variety farther away. Darwin regarded such situations as examples of evolution in the making. The bird varieties were diverging branches of a common ancestry, each an incipient species.

Some biologists, however, could not accept the idea that many tiny changes could produce the vast differences that separate, say, a horse from a cow or, for that matter, an insect from a human. They felt there had to be some other mechanism that would produce big changes in a short time. In 1900 it looked as if a mechanism for big changes had been discovered. Gregor Mendel's pioneering experiments in breeding pea plants—performed forty years earlier but ignored, forgotten, and rediscovered—showed that organisms carry their inherited traits in discrete particles (later called genes). Mendel's pea plants were either tall or short, the seeds wrinkled or smooth, the blossoms purple or white—with each trait governed by a distinct gene. As geneticists later learned, most traits are governed by many genes, and their patterns of inheritance are much more complex than those in Mendel's pea plants. But in 1900 this was not fully appreciated. To the geneticists rediscovering Mendel and confirming his experiments, it looked as if they had, at last, found the mechanism by which sudden evolutionary jumps could occur.

Thus, biologists of the early twentieth century were divided into two camps. There were the geneticists, working mainly in the laboratory, who saw only evidence for sudden and discontinuous change. And there were the naturalists, working out in the field, who saw only gradualism in the observable world.

Slowly the rift was healed by the modern synthesis. This came largely as the result of geneticists realizing that most hereditary traits are governed by many genes working in concert. The modern synthesis was forged by many scientists, the best known of whom was probably Theodosius Dobzhansky, a Russian-born American geneticist, with his experiments on fruit flies. With Dobzhansky's work it became clear that genetic changes were usually of small effect and could well accord with Darwinian gradualism. Moreover, Dobzhansky's breeding of fruit flies under various environmental conditions in the laboratory seemed to confirm Darwin's emphasis on natural selection. The fruit flies actually evolved in the laboratory to become better adapted to various artificial environments.

As part of this emerging new view of evolution, Dobzhansky emphasized the need for what he called isolating mechanisms. He recognized that in the wild, a new species could not emerge from an old one if its early members (the variants on their way to becoming a new species) could breed with the parent stock. The novel features of the variants would either be swamped by the larger, nonvariant population, or they would be spread throughout the existing species, causing the entire population to evolve slightly. No new species could then split off. Exactly this process sometimes happens in nature, but biologists realized that if it were the only method of evolution, there would be no splitting off of new species, and the diversity of life forms could not have increased over geologic time.

If part of a species population is to become a new species, Dobzhansky argued, it must be isolated from the rest of the species. A river, mountain range, or some other geographic feature must prevent this small variant group from breeding with its original stock. Eventually, the isolated population will then accumulate enough changes that even if it somehow migrates back into its ancestral area, it will be too different to interbreed with the stock from which it came.

Over the years, several scientists made other contributions to the modern synthesis. In 1944 the "synthesis" took a big step forward when George Gaylord Simpson, an American paleontologist, published a book called *Tempo and Mode in Evolution*, linking Darwin's theory and modern genetics to the fossil record. He was trying to cope with one of the biggest embarrassments facing Darwinian gradualism—the fact that it was very hard to find fossils of intermediate species bridging the evolutionary gap between major groups of animals. If gradualism were correct, one could hope to find the fossil bones of creatures that were transitional between, say, the early rodent-like mammals and a giraffe, or the species intermediate between a primitive fish and the first land-dwelling vertebrate.

Darwin was aware of these gaps in the fossil record, but assured his followers that further hunting would eventually turn up the creatures that would fill them. But as Simpson knew, a century of fossil hunting after Darwin had turned up precious few such "missing links." To be sure, however, some striking examples were found.

The best known is *Archaeopteryx*, a fossil species from the age of Jurassic dinosaurs, that was thought on the basis of its skeleton to be a small dinosaur. But a remarkably preserved specimen was found surrounded by the imprint of feathers. It had teeth and its front legs had claws, but they also had long, wing-type feathers. The specimen neatly occupied a gap between reptiles and birds. Although Simpson took the few gap-fillers like *Archaeopteryx* to vindicate gradualism, he also argued that in perhaps 10 percent of cases the gaps meant that evolution had suddenly surged, spawning a new species within only a few generations and in such an isolated region that there would be no hope of finding any surviving fossils. Simpson called this "quantum evolution" (not to be confused with the quantum of atomic physics).

The idea of evolution making a sudden leap, however, did not sit well with other biologists, and in the 1950s Simpson modified his ideas to better fit the prevailing belief in gradualism. The modern synthesis, once a fairly pluralistic collection of evolutionary ideas, became narrower and more dogmatic in its insistence on Darwinian gradualism and natural selection.

The orthodoxy of gradualism held until the 1970s, when two paleontologists—Stephen Jay Gould and Niles Eldredge—reintroduced the idea of sudden surges of evolution in a theory they called "punctuated equilibria."

See EVOLUTION; NATURAL SELECTION; DARWIN; MENDEL; GENES, HOW THEY WORK; PUNCTUATED EQUILIBRIA.

MOLECULE—The smallest unit of a chemical compound that is capable of a stable, independent existence. Most molecules are made of several atoms bound together. A molecule of hydrogen, for example, consists of two hydrogen atoms bound by the sharing of each other's lone electron. The sharing gives each atom a stable configuration of two electrons (covalent bonding). A molecule of water consists of two hydrogen atoms bound to one oxygen atom. Again, electrons are shared, giving each atom a stable number of electrons. Some molecules, such as those of water, are quite small. Others may contain thousands of atoms bound into complex patterns. Typically, organic molecules such as proteins have hundreds or thousands of atoms.

See ATOM; CHEMICAL BONDING.

MONOCLONAL ANTIBODY—A special type of antibody, manufactured in a laboratory in a tissue called a hybridoma, capable of binding to any specific protein molecule for purposes of diagnosis or treatment.

See ANTIBODY; HYBRIDOMA.

MOON, ORIGIN OF—
See SOLAR SYSTEM, ORIGIN OF.

MORGAN, THOMAS HUNT, 1866–1945, American biologist. See DNA, HOW ITS ROLE WAS DISCOVERED.

MUTATION—A mutation is any change in the genetic message. It may result from a toxic chemical or particle of radiation that hits the atoms making up a gene, converting one configuration of atoms into another. Mutations can also be caused by errors within the cell nucleus as a cell divides, with the result that a gene is miscopied. Still another cause is a chromosome breakage followed by incorrect reassembly of the broken pieces. As the result of such an event, half of one gene may get spliced onto the end of another.

Usually mutations, which occur at random within genes, are harmful or of no consequence. Individuals carrying such mutations survive poorly if at all, and usually don't reproduce. The rare favorable mutation, however, can cause an organism to thrive unusually well, increasing the chances that the mutation will be passed on to many descendants. All living organisms are mutants, inheritors of many fortuitous alterations in the genetic messages carried in the genes of our ancestors. Without mutations, in fact, evolution would never have taken place. Mutations may occur in any cell, but only those that occur in the cells that produce ova or sperm have consequences for the future generations of a species.

See GENES, HOW THEY WORK; EVOLUTION; NATURAL SELECTION.

NATURAL—All good languages change with time, but there seems something perverse about the way "natural" gets misused in today's culture. In advertising for processed foods, for example, it is as meaningless as "new and improved." Yet there is a certain inescapability in this for, in a strictly scientific context, anything is natural and everything is natural. Nature, after all, is what is, and human beings and their culture are as valid a part of nature as any other objects. Thus, a human apartment building is no less natural than a beehive. Even great, environment-modifying projects such as dams are no less natural than beaver dams that destroy forests. All human behavior is subject to "natural law," which, in fact, is impossible to break (unless you believe in miracles). Therefore, while it may be irksome when advertisers give the "natural" label to machine-planted, mechanically harvested, chemically processed, physically cooked, synthetic-wrapped food products, it is valid and, ultimately, meaningless.

NATURAL SELECTION—A mechanism by which evolution works, at least some of the time. Darwin considered natural selection to be the main process driving evolution. Some of today's theorists think other mechanisms are more powerful. The basic concept in the theory of natural selection is that any physical or behavioral trait that improves an individual's chances of producing viable offspring will show up in even greater prevalence among members of the next generation. As a result, a trait that appeared as a mutation in only one individual could eventually become standard within a species.

The theory has a certain inescapability about it. Any genetically determined trait that causes an individual to outreproduce its brethren can hardly fail to become more and more widespread within the species. It was this inherent plausibility that led so many biologists in Darwin's day to immediately accept his theory. It is still plausible that the process of evolution works according to the theory. What is in dispute now is whether natural selection is the sole or even main engine pushing evolutionary change, especially such major steps in evolution as the formation of new species.

See FITNESS; DARWIN; PUNCTUATED EQUILIBRIUM; GENETIC DRIFT; EVOLUTION.

NEBULA—This is a term in astronomy for any hazy patch of sky or similarly nebulous object in the universe. As telescopes improved, it became clear that some nebulae were actually huge clusters of distant stars. Henceforth, these nebulae were called galaxies. However, there are still nebulae, some of which are believed to be vast clouds of gas and dust drifting about in space. There are also stellar nebulae consisting of a star surrounded by a dense cloud of dust and gas in which, it is believed, planets are forming just as did our solar system.

See UNIVERSE, STRUCTURE OF; GALAXY; SOLAR SYSTEM, ORIGIN OF; LAPLACE (an early advocate of the nebular theory of planet formation).

NEURON—A nerve cell, whether in the brain or elsewhere. A typical neuron consists of a roundish main body from which tentacles project. The shorter tentacles are tree-shaped, and are accordingly called dendrites. There is one long tentacle called an axon that may stretch several inches, and in some cases several feet. The axon may be branched at its far end. Nerve signals travel through the neuron as electrical impulses that pass down the axon and then jump a gap from a branch of the axon to the dendrite of another neuron. The gap is called a synapse, and the signal crosses by means of chemicals released from the tip of the axon. The chemicals are called neurotransmitters.

NEUROTRANSMITTER—Any of several chemicals produced in nerve cells and released in bursts to carry signals to other nerve cells. The best known are acetylcholine, serotonin, epinephrine (also called adrenalin), and norepinephrine (also called noradrenalin). Many brain disorders, including certain psychiatric illnesses, are thought to be the result of defective production of certain neurotransmitters, which results in faulty transmission of information from one nerve cell to another.

See NEURON.

NEUTRINO—At any given moment there are about 100 million of these subatomic particles penetrating your body as they zoom in from outer space. They don't take long to go through, since they travel at the speed of light. And since they are so small and have no mass, they pass right through without hitting any of the particles that make up the atoms of your body. In fact, neutrinos pass through the entire Earth at the same speed that they pass through the body, slipping past all the atoms on the way with never an encounter. There are so many neutrinos whizzing about in every direction that, in a sense, the universe is flooded with them. Besides zipping about on the loose, there are also neutrinos inside atoms. They don't exist as such, but do form under special circumstances when atoms are forcibly broken up, such as in a particle accelerator. When this happens the neutrinos come from the breakdown of a neutron. In this sense, there is a neutrino inside a neutron, along with other particles.

Its name makes a neutrino sound like an Italian particle, and in one way it is. The name was invented by the great Italian atomic physicist Enrico Fermi for a particle that was neutral in charge and had no mass. In Italian it means "little, neutral one."

See ATOMIC STRUCTURE; NEUTRON.

NEUTRON—One of the three main particles of which all atoms except hydrogen are made. Hydrogen consists of a nucleus of one proton with one electron orbiting around it. The atoms of all other elements have protons and neutrons in their nuclei, with the total number of these particles roughly proportionate to the weight of the atom. Like protons, neutrons are made of quarks—three of them, designated as two "down" quarks and one "up" quark. (Down and up, like terms for other types of quarks, are whimsical terms meant only to imply opposite states.) Neutrons have no electrical charge. Within atoms they are stable, but isolated they disintegrate spontaneously with a half-life of about 15 minutes. A decaying neutron will first split into a proton and a particle called the W⁻, which will then further split into an electron and a neutrino.

See ATOMIC STRUCTURE; PROTON; QUARK; ELECTRON; NEUTRINO; HALF-LIFE.

NEUTRON STAR—See PULSAR.

NEWTON, ISAAC 1642–1727, English physicist—One day young Isaac Newton was sitting in his mother's orchard when he happened to see an apple fall. The observation launched Newton on a train of thought that led to one of the greatest scientific achievements of all time, the law of universal gravitation. This is the idea that all objects exert a gravitational pull, and that the pull increases with the mass of the object but decreases with distance from the object.

Familiar storybook episodes like that of Newton in the orchard are supposed to be of doubtful authenticity, but it appears that this experience really did happen to Newton, except that, contrary to most popular versions, the apple did not hit him on the head other than figuratively. Newton was only twenty-three at the time, and he had, within the previous few months, already formulated his famous laws of motion, the three deceptively simple statements that govern the motion of all bodies from apples to galaxies. One of them, for example, describes the action-reaction principle upon which rockets work. The falling apple triggered not only the discovery of the laws of gravitation, but led young Newton to invent the calculus and, almost as a sidelight, to formulate the mathematical laws governing the way all colors combine to make up white light.

Newton accomplished all this before his twenty-fifth birthday, during one of the two most spectacular periods of scientific productivity in all of history. The second such period also involved Newton. It came twenty-two years later when he would write up his theories. Neither period lasted more than two years, but during them Newton did all the work that would establish him as possibly the greatest scientific genius of all time. Lots of scientists have been ranked among the greatest, but only Newton has consistently earned the highest praises from his own time three centuries ago to the present. No less a figure than Einstein, who found Newton's laws to be invalid in the subatomic realm, considered his predecessor to have been perhaps the greatest scientific mind of all time.

However, during Newton's other eighty-one years of life, he did comparatively little. In fact, there were nearly two decades between his two great bursts of productivity in which he did nothing very remarkable. Aside from his scientific achievements, Newton is mainly remembered for being a moody, tortured, often disagreeable person who was always getting into arguments. He never married but, as the historians say, remained devoted to his mother. He was a tension-ridden man who suffered at least two nervous breakdowns. One of them, upon the death of his mother, sent him into morbid seclusion for six years, during which he saw few people and seldom ventured from his house.

Isaac Newton was born on Christmas day in 1642, the year Galileo died. After living his first eighteen years at home in the little hamlet of Woolsthorpe, Newton was sent to Cambridge University. He studied under a mathematics professor named Isaac Barrow, who ought to be considered the patron saint of teachers everywhere. Barrow is one of two people—Edmund Halley, of the comet, is the other—without whom Newton's genius might never have emerged from his moody soul. Barrow recognized Newton's genius, encouraged him to develop his abilities, and showed him the way. In college, Newton discovered the binomial theorem, a trick of algebra that simplifies certain kinds of calculations. This alone would have secured him a place in the history of mathematics.

Before Newton could finish college, the Great Plague, which had already killed a third of London's population, spread to Cambridge and the university was closed. The year was 1665, and Newton went back to Woolsthorpe to live with his mother. Isolated once more from the intellectual world, he devoted himself to scientific experiments and meditation. The next two years were to be the most productive of his life. Like Copernicus and Galileo, Newton was fascinated by the movement of celestial bodies, including the Earth. His two great predecessors had established that the Earth moves about the sun, but no one yet knew why. What was it that kept the motion of the moon and the planets so orderly? When Newton grappled with the problem, he found the mathematics of his day inadequate to answer it. There was no way to deal with motion, with numbers that changed continuously as a phenomenon progressed. So Newton invented a way. He called it fluxions; today we call it differential calculus.

Having invented the key, Newton unlocked the universe. In particular, he discovered the law of universal gravitation, of the phenomenon that holds the universe together, binding the moon to the Earth and the Earth-moon system to the sun. And, for that matter, the sun to the galaxy, although Newton never knew about galaxies. There was nothing new in noticing that objects—like the apple in his mother's orchard—fall to the ground; everybody knew the Earth exerted some kind of pull. Newton's contribution was the idea that the apple also exerted a pull on the Earth. (The Earth does move ever so slightly toward a falling apple.) In fact, Newton asserted, every object in the universe exerts a pull on every other object. This assertion alone would have been a remarkable enough insight. But Newton did more. He used his calculus and proved, on the basis of the known movements and distances of various celestial bodies, that the law of universal gravitation was true.

It was a stunning intellectual achievement but, as befit his secretive personality, Newton sat on his discovery for twenty-two years before publishing it. Likewise, he told no one about his invention of the calculus or of his discoveries about the colors of light, which also came during this time. It was in this same phenomenally productive two years that Newton formulated his famous three laws of motion. They state that: (1) objects in motion or at rest tend to stay that way unless acted upon by an outside force; (2) force is something that produces acceleration in a body (acceleration is an increase in the rate of motion of a body, and it can make things move faster or slower); and (3) that for every action, there is an equal and opposite reaction.

After two years the plague threat subsided at Cambridge, and Newton returned to finish his studies. He received a master's degree and was made a fellow of

Cambridge's Trinity College. Within two more years Newton's old mentor, Isaac Barrow, stepped down from his professorial post, yielding it to Newton, who was just twenty-seven. For the next twenty-seven years Newton lived as a bachelor in dingy, unkempt rooms near the campus and taught mathematics. Except for a paper on his experiments with light and color, this was a most unremarkable period in Newton's life. And that one paper was much to blame for this.

Newton wrote the paper at Cambridge, laying out his conclusion—which was based on experiments with a prism—that ordinary "white" light is a mixture of all colors of the spectrum. Though he had mathematical calculations—dealing with the bending o light as it passes through the prism—to back up his claim, Newton was attacked by the scientists of his day. Not only had he repudiated the orthodox view of the nature of color, he had made the mistake of asserting that "the proper method for enquiring after the properties of things is to deduce them from experiments." That, too, ran counter to orthodoxy, which preferred simply either to quote Aristotle or to invent explanations from a combination of imagination and logic. Gentlemen simply did not dirty their hands with experimentation.

Prominent scientists such as Christian Huygens and Robert Hooke attacked Newton so severely that he resolved never again to publish. Newton wrote that he was "persecuted" by his critics, and that the incident left him so bitter he even lost his "affection" for science. Indeed, it appears that he engaged in almost no science for the remainder of his life. Events, however, would eventually force his hand.

By the 1680s other scientists had begun to converge on the idea of universal gravitation, but without mathematical proof. One of those working on the problem was Edmund Halley, the astronomer, who calculated the return of the comet named for him. Halley sought Newton because of his reputation as a mathematician, and learned that he had already solved the problem, along with many related problems. Halley understood the significance of

Newton's work and urged him to publish it. Newton said no, but Halley kept insisting. Newton, however, stubbornly refused; he was a painfully private man, fearful of public attention, and he dared not risk the agony of an open controversy. Halley continued to press Newton, talking excitedly about how his discoveries could be developed and used. Halley even negotiated with the Royal Society, getting them to pay the publishing costs. Gradually Newton found his old enthusiasm returning and, at last, agreed to write a book.

Newton set to work, writing the book that his entire life in science had prepared him to write. He wrote furiously, missing meals, sleeping little. For eighteen months he scribbled—long, convoluted text passages, always in scholarly Latin, relieved by some of the most elegant mathematics the human mind has ever conceived. The result would be one of the most celebrated books in science, *Philosophiae Naturalis Principia Mathematica*. In English, its title means Mathematical Principles of Natural Philosophy, but it is usually known simply as the *Principia*. When Newton was two-thirds of the way through with the work, Hooke got wind of it and publicly claimed that it was he, not Newton, who had been first with the law of universal gravitation. The truth is that Newton thought of it first but kept quiet while Hooke, who thought of it later, announced it first. Before he had even published, Newton was again tormented by controversy. He threatened to stop writing, but again Halley intervened and talked Newton into finishing the book. Just as the completed manuscript was going to the printer, the Royal Society shortsightedly backed out of its arrangement to finance it. Once more Halley stepped in. Though not a wealthy man, he had more money than Newton, and he dipped into his own pocket to pay the printer. He even put aside his own work to supervise the printing, ensuring its accuracy.

The *Principia* was published in 1687. It is 250,000 words long—two and a half times the length of a typical novel—and all of it is in the deadliest of classical Latin. It is perhaps the greatest single book in science and quite likely the least read. In its day and with its deliberately narrow audience, it was an earthquake.

The *Principia* is in three parts, or "books." The first deals with the motion of bodies in free space. After explaining differential calculus, upon which most of the rest of the *Principia* depends, Newton set forth his laws of motion. The second book deals with motion in a resisting medium such as water, and with the motion of fluids themselves. It explains how to determine the speed of sound and how to describe waves mathematically, and sets forth other principles of the fields that would some day be called hydrodynamics and hydrostatics. In the second "book" Newton disproved the prevailing theory of how the planets moved, which had to do with Descartes's theory of vortexes. (Descartes claimed that the planets swirled about the sun like objects caught in a whirlpool. Newton showed that a "resisting medium," such as would be required for a whirlpool, would make the planets behave in ways quite unlike the ways in which they really behave. Newton, one must remember, had been reviled for insisting that it was necessary to have some physical evidence before hatching a theory.)

The third book of the *Principia* is entitled "The System of the World." Only from Newton could such a title not seem hyperbole. This book begins with the bold assertion that the laws of nature are the same everywhere. "Like effects in nature are produced by like causes, as breathing in man and in beast, the fall of stones in Europe and in America, the light of the kitchen fire and of the sun, the reflection of light on the earth and on the planets." Elementary as this notion may seem today, it was revolutionary in its time. It meant that experiments done on Earth might explain phenomena far out in space. At a stroke, the mysteries of the universe were made amenable to direct study. Book three goes on to explain the motion of the planets and moons, and to give methods for finding the masses of the sun and planets. Remarkably, Newton's estimate of the Earth's density (between five and six times that of water) was very close to today's estimate of 5.5. Newton also explained tidal cycles, including the separate roles played by the sun and moon, and explained the precession of the equinoxes. Beyond this, he offered the theory that comets were not supernatural omens but physical objects, obeying the laws of gravity like any other objects, and established that they were traveling in very large and extremely elliptical orbits about the sun. He observed that the Earth bulges at the equator and is flattened at the poles—a consequence of its rotating on its axis.

If the question had to do with an object in motion, the *Principia* dealt with it. Whether the object was a planet or a pebble, Newton's genius lay in perceiving that it obeyed a common set of laws. (It should be noted here that twentieth century scientists would find a realm where Newtonian physics does not apply—the movement of particles inside the atom.) Great as Newton's achievements were—and despite Newton's personal isolation—they were made possible by the pioneering work of predecessors such as Copernicus, Kepler, and Galileo. Newton himself acknowledged his debt with the now classic remark, "If I have seen farther than other men, it is by standing on the shoulders of giants."

Still, the *Principia* did not become a runaway best seller. It caught on slowly, taking ten to twenty years to establish itself as the vehicle of the new orthodoxy. Some of Newton's fears about stirring up controversy were borne out. Hooke, as we have seen, claimed that he had been first with the theory of universal gravitation and kept up his attacks on Newton. Gottfried Leibniz, a mathematician, claimed priority in the invention of the calculus. Again, the fact is that Newton thought of it first but published it second.

After his eighteen-month burst of productivity, and suffering from personal exhaustion, Newton lapsed into another period of relative scientific inactivity. For the next forty years, until his death, he essentially engaged in no more science. But as the importance of the *Principia* gradually sank in, honors flowed to Newton in abundance.

He was, in fact, fully appreciated in his own time. Queen Anne appointed him Master of the Mint and knighted him. He was elected president of the Royal Society in 1703 and reelected every year thereafter until his death in 1727.

For all the acclaim, Newton remained a lonely man to the end. Summarizing his own life, he wrote, "I do not know what I may appear to the world, but to myself I seem to have been merely a child playing on the seashore, diverting myself in now and then finding a pebble more smooth or a shell more beautiful than others, whilst before me the great ocean of Truth lay all undiscovered."

See GRAVITY; KEPLER; COPERNICUS; GALILEO; HOOKE.

NIRENBERG, MARSHALL, 1927– , American biochemist. See GENES, HOW THEY WORK.

NOBLE GAS—A chemical element that is inert, never taking part in reactions with other elements. There are six such elements. They are "chemical snobs," and for this reason got their name. The elements are helium, neon, argon, krypton, xenon, and radon. All normally exist as gases in the atmosphere. (Incidentally, since krypton is inert, it could never form the compound kryptonite. Superman is safe.)

NONSENSE DNA—This is a scientific *gxrwksy* colloquialism for the vast amount *rtswqplobde* of DNA in cells that seems to carry no discernible genetic message. It is, in effect, genetic gibberish that interrupts genetic messages just as do the italicized "words" in the previous sentence. Besides occurring within genes, sequences of nonsense DNA may also occur between genes. When the message in a gene is transcribed from DNA into RNA (which carries the genetic code from the nucleus to the cell's protein-making apparatus), it is edited so that the nonsense passages are removed and the sensible parts spliced into one uninterrupted sequence. The gibberish sequences are sometimes called introns, or intervening sequences, while the message parts of the gene are called exons, or expressed sequences. There is evidence that in some genes the nonsense portions may serve as regulators of gene activity. Also, it is clear that some genes have intervening sequences that, once edited out during RNA transcription, become part of another gene.

See DNA; GENES, HOW THEY WORK; GENETIC CODE.

NUCLEAR WINTER—We now have yet another reason to prevent nuclear war. As if the blast effects, firestorms, and radiation weren't enough, it is now believed that a nuclear war might alter the Earth's climate enough to threaten to destroy life all over the planet. The dust and smoke thrust into the upper atmosphere by even a "modest" nuclear exchange would so darken the sky that much less sunlight could reach the ground than now does. The particles would remain aloft for weeks or months, depriving the ground of the solar heat that ordinarily makes the difference between winter and summer. Winter would grip the entire planet and could destroy plant life in the tropics and agriculture throughout the northern hemisphere.

NUCLEIC ACID—Nucleic acids are the long-chain polymer molecules that store and transmit the genetic code. One type is deoxyribonucleic acid. Known as DNA for short, this chemical is what genes are made of. Another type of nucleic acid is ribonucleic acid, or RNA. There are several kinds of RNA, including one that transcribes the DNA message and carries it to the cell's protein-making machine, the ribosome. This type of nucleic acid is called messenger RNA, or mRNA. A different kind of RNA brings amino acids (the small molecules of which proteins are made) to the ribosomes for use in building various proteins. This nucleic acid is called transfer RNA, or tRNA.

Nucleic acids are among the most remarkable of substances, but some advertising copywriters seem to misunderstand the reason why. Several brands of shampoo and cosmetics are advertised as containing nucleic acids, as if these molecules were some kind of wonder ingredients. Actually, they do nothing for the chemistry of cosmetics. Nucleic acids are in all cells and everybody who eats meat or vegetables takes in a dose of nucleic acids in every cell of the food. Within each cell nucleic acids carry the hereditary information of the species and govern the life of the cell. But, when eaten, they are broken down by the processes of digestion and have no special effect on the eater.

See GENES, HOW THEY WORK; GENETIC CODE.

NUCLEOTIDES—These are the basic subunits of which DNA and RNA are made. Each of the two complementary strands of DNA consists of a chain of thousands of linked nucleotides. Each nucleotide has three parts: a sugar (deoxyribose in the case of DNA and ribose in the case of RNA), a phosphate, and a base (one of the four "letters" of the genetic alphabet—adenine, thymine, guanine, and cytosine. The sugar and phosphate are linked together to form the backbone of each DNA chain. The bases stick out perpendicularly from each chain, and are linked to their complementary bases on the opposite chain, thus holding the two chains of DNA to one another in a "double helix" configuration. The sequence of bases, or letters of the genetic alphabet, specifies the genetic message that is encoded in a nucleotide.

See DOUBLE HELIX; GENES, HOW THEY WORK; GENETIC CODE.

NUCLEUS (ATOMIC)—In the center of every atom is the nucleus, which is made of protons and, except for hydrogen, neutrons as well. For every proton in the nucleus there is usually one electron orbiting the nucleus. Since protons and neutrons each weigh about 1,800 times as much as electrons, virtually all the weight of an atom (and hence virtually all the weight of all matter) is in the atomic nucleus. The amazing thing about the nucleus is that it is able to hold together several protons which, because they carry the same positive electrical charge, normally repel one another (according to the law of electromagnetism by which opposites attract and likes repel.) They are bound by the most powerful force known, the so-called strong nuclear force. The strong force is thirty-nine orders of magnitude stronger than gravity. That's a 1 followed by thirty-nine zeros. But whereas gravity reaches out infinitely through space, the strong force extends no farther than the atomic nucleus.

See ATOM; ATOMIC STRUCTURE; STRONG FORCE.

NUCLEUS (CELL)—Every living cell (except those of bacteria and blue-green algae) contains a nucleus, a porous bag holding the genes, which are assembled single file on chromosomes. Genes, made of DNA, always stay inside the nucleus. Their coded messages reach the rest of the cell via a messenger molecule called messenger RNA, or mRNA. The RNA is produced in a part of the nucleus called the nucleolus, a bag within the nuclear bag. To send its message, the double helix of the DNA unzips, exposing a sequence of subunits—the four "bases" that make up the genetic alphabet—so that a complementary, messenger RNA sequence can be assembled. The messenger RNA then leaves the nucleus and drifts in the outer part of the cell, where ribosomes, the protein factories, "read" the code being carried by the messenger RNA and assemble the protein that it specifies.

See DNA; RNA; GENES, HOW THEY WORK; DOUBLE HELIX; NUCLEOTIDES; MITOSIS (the process by which chromosomes duplicate themselves so as to form two new nuclei prior to cell division.)

OCHOA, SEVERO, 1905– , Spanish-born American. See GENES, HOW THEY WORK.

OERSTED, HANS CHRISTIAN, 1777–1851, Danish physicist. See FARADAY.

OHM, GEORG, 1787–1854, German physicist. See CAVENDISH.

ONCOGENE—The most important development in cancer research in recent years has been the finding that every cell of almost every species contains genes that can become cancer-causing with only the slightest alteration. These genes are now known as proto-oncogenes. After the alteration they become oncogenes, *onco* being derived from the Greek word for tumor. The alteration may involve as little as the equivalent of one letter in a paragraph of several hundred words. Even so, it is enough to cause the gene to produce a different protein or to speed up its production of a normal protein. Either way, the cell begins proliferating without hindrance, as opposed to normal cells, which stop dividing in response to certain signals.

At first no one knew that human cells contained proto-oncogenes. Cancer genes seemed to exist only in certain viruses that caused cancer in animals. When most viruses infect a cell, their genes "hijack" the cells's protein-making machinery and use it to make new viruses. Eventually this can kill the host cell. The new viruses then burst out and invade new cells. Cancer viruses are of a sneakier type. Instead of directly commandeering the cell's metabolic machinery, these viruses splice their genes directly into the cell's DNA, where they behave just like native genes. Instead of killing its host cell and moving on, like the transient, ordinary virus, a cancer virus becomes a permanent resident of the cell. When the cell undergoes normal cell division, the inserted viral oncogene is replicated, so that one copy goes into each daughter cell. It is not clear how they do it, but virus oncogenes cause cells to ignore the signals that would normally halt their proliferation. The cells keep multiplying, producing a tumor.

When molecular biologists decoded the genetic messages on viral oncogenes, they turned out to be extraordinarily close to the messages already encoded in certain genes of almost every living organism. Often they differed by only one base or "letter" in the genetic message out of thousands. A mutation at that base, converting it into a different base, could convert the normal gene into an oncogene. Thus these animal genes came to be called proto-oncogenes.

It is almost certainly not coincidence that cancer viruses have genes so closely resembling the genes of animals. In fact, it seems likely that the viruses are escaped groups of mutant normal genes.

See DNA; GENES, HOW THEY WORK; GENETIC CODE; VIRUS.

OPEN UNIVERSE—See CLOSED UNIVERSE.

ORGAN—A significant part of a plant or animal that constitutes a structural and functional unit. An organ may be composed of several kinds of tissues. It is easy enough to see that hearts and kidneys are organs. But in fact, bones are also organs, as is skin. The distinction of organs from tissues is not always clear-cut. Blood, for example, is often sp̣ken of as both a tissue and an organ, the latter especially if one is thinking of all of the body's blood as a unit. Muscles can be considered in much the same way. In plants, the term organ is broader still. A leaf is an organ, as is a root or stem.

ORGANELLE—These "little organs," as the name implies, are to cells what organs are to whole individuals. Among the organelles of cells are the nucleus (which holds the genes), the mitochondria (which extract chemical energy from food), and the ribosomes (which "read" genetic messages and assemble protein molecules in response to them). Plant cells also have chloroplasts (which carry out photosynthesis).

See MITOCHONDRION; RIBOSOME; NUCLEUS; CHLOROPLAST.

ORGANIC—This is another word, like "natural," that has been much abused by advertising copywriters. It is also about as meaningless a word as "natural." Historically, "organic" referred to any unique property of living organisms, such as the ability to reproduce, or irritability.

Nonliving systems and their properties were said to be
inorganic. Today that distinction is seldom used, and
about the only scientific context in which the term
"organic" is still used is the branch of chemistry that deals
with carbon compounds. Organic chemistry, as the disci-
pline is called, originally dealt with the compounds synthesized in living organisms, but it soon became clear that
these substances could be made in the laboratory by entirely nonliving methods (if you don't count the chemist).
Thus, organic chemistry deals with practically all carbon-
based substances. Since these include the vast majority of
synthetic pesticides, it makes little sense to reserve the
label "organic" for foods grown without such chemicals.

OSMOSIS—This is the physical phenomenon that plant
roots use to get water from the soil and that blood uses to
get oxygen from the lungs. In fact, osmosis occurs
throughout plants and animals as a means of moving various substances from one place to another. It is not the
only means for doing this, but it is a common one, and it
works by a process that is entirely physical, requiring no
overt control by the organism.

In osmosis, molecules of liquids or dissolved substances
pass from one side of a semi-permeable membrane to the
other. Molecules that are small enough to pass through
openings in the membrane will move in both directions
randomly, but if there are more of these molecules on one
side than on the other, chance alone will dictate that more
molecules from the richer side will pass through to the
poorer side than the other way around. Eventually the
concentration of molecules will reach an equilibrium, so
that as many molecules move in one direction as move in
the other.

If sugar water in a membrane bag is put in a bowl of
pure water, the concentration of water molecules outside
the bag will be higher than inside it. If the membrane will
let water molecules through, but not the larger sugar molecules, more water molecules will enter the bag than leave
it, and the bag will gradually swell up as the water pres-

sure inside it increases. This is precisely the mechanism by which roots take in water. In fact, all cells are surrounded by a membrane, and as the water pressure inside the cell increases, the more turgid the cell becomes. This is why plants wilt in dry soil but stiffen up when there is water to swell their cells.

If the membrane separates mixtures of different kinds of molecules, all of which can pass through it, the passage of each kind will be independent of the others. Each will "try" to reach its own equilibrium. Thus, in the bloodstream that passes through the lungs, the higher concentration of waste carbon dioxide in the blood will diffuse into the air as the oxygen from the air diffuses into the blood.

See MEMBRANE.

OVUM—The unfertilized egg cell produced by the female of all sexually reproducing plants and animals. The ovum contains half the chromosomes appropriate to cells of the species, and when fertilized by the male's sperm, bearing a complementary half set of chromosomes, becomes the first cell of a new embryo. In most species the ovum is thousands of times larger than a sperm, mainly because it contains all the cellular machinery needed to be a complete living cell, as well as the nutrient supply (or yolk) needed until the embryo is developed enough to draw on outside sources. Bird eggs are the largest of all cells. In birds and reptiles, the yolk is very large and must meet all of the nutritional needs of the embryo until it hatches. In the eggs of mammals, the yolk is quite small because the embryo soon grows a placenta to tap the mother's bloodstream.

See CELL.

PALEOMAGNETISM—Some day, all our magnetic compasses will be wrong. Instead of pointing to the Earth's north pole, they will point to the south pole. This is because the Earth's magnetic field reverses itself every few tens of thousands of years. Nobody knows why it does this, but there is no question it has happened dozens of times in the past, because the effects of reversed magnetic fields have been found in ancient rocks. These effects were caused by paleomagnetism, or "old magnetism."

Paleomagnetic effects are most noticeable in igneous rocks, the kind that form as molten lava cools. When the rock is in the liquid state, atoms within it are able to move easily. Since each atom is itself a tiny magnet, there is a tendency for the atoms to align themselves with the polarity of the Earth's magnetic field. Once the lava cools, the atoms cannot move, and the polarity of the rocks is frozen into position. The next time the Earth's magnetic polarity reverses, the polarity of the old rock will stay the same, while the new rock that forms on top of it will acquire the reversed polarity.

Since the Earth's magnetic field affects the entire planet, geologists have been able to use the sequence of magnetic reversals in the Earth's history as a way of correlating rock layers in one part of the world with those in another. In this way, it is often possible to assign ages to rocks, a practice known as paleomagnetic dating.

See PLATE TECTONICS (a field in which paleomagnetism contributes important evidence).

PANGAEA—The name of the one supercontinent that existed millions of years ago, before it broke into several continents, which then drifted to their present positions.

See GONDWANA; PLATE TECTONICS.

PARACELSUS, 1493–1541, German physician and chemist—In all the history of medicine, there are few more colorful characters than Paracelsus, the man who launched the medical wing of the Renaissance by rejecting some two thousand years of astrology, magic, and herbalism and declaring that diseases have natural causes and can often be treated with chemically prepared drugs. He rejected the ancient doctrine of humors, fluctuations of which were supposed to cause disease. He recognized that silicosis and tuberculosis were not punishments for sin but occupational hazards of miners resulting from the inhalation of noxious substances. He also recognized numerous neurological and psychiatric disorders as having natural, not supernatural, causes. And he claimed wounds would heal themselves if doctors would confine their actions to preventing infection. For his heretical teachings, Paracelsus was branded the medical Luther.

His full mane was Philippus Aureolus Theophrastus Bombast von Hohenheim, but he so loudly proclaimed his own superiority to all other physicians that he renamed himself Paracelsus, which means beyond, or greater than, Celsus, the great first-century Roman physician. So angrily and vociferously did Paracelsus rail against the superstitious medical orthodoxy of his time that his surname, Bombast, became the root of the adjective "bombastic."

Paracelsus was born in what is now Switzerland, the son of a modest country doctor who was also a chemist. He lived a peripatetic life. At the age of fourteen he left home and joined the swelling ranks of young people who at that time traveled all over Europe seeking out famous professors. In five years, he attended the universities of Basel, Tubingen, Vienna, Wittenberg, Leipzig, Heidelberg, and Cologne. None satisfied him. In a remark typical of his vituperative nature, he asked how "the high colleges managed to produce so many high asses." He claimed he got a better medical education by consulting "old wives, gypsies, sorcerers, wandering tribes," and other sources of practical knowledge.

Even after taking his medical degree, Paracelsus kept up his travels, always seeking new medical knowledge. He spent time in almost every European country, including England, Ireland, and Scotland. He visited Russia, was imprisoned there, escaped to Lithuania, and eventually wound up as an Army surgeon in Italy. Later he went to Egypt, Arabia, Palestine, and Constantinople. In each place, he said, he consulted with leading doctors and with alchemists, whose new and curious concoctions Paracelsus studied for any potential medicinal value they might have.

When he returned to the Germanic regions from which he had come, he found that his fame had preceeded him. He quickly gained a position at the University of Basel and, much to the ire of the administrators, invited the public to attend his medical lectures. He railed against the worthless potions and herbs of the orthodox practitioners, creating such public doubt about doctors and apothecaries that they eventually threatened his life. He had to flee in the dead of night and, for the next eight years, he wandered Europe, staying with various friends and working on his manuscripts. When they were published, they won such acclaim that Paracelsus re-emerged into public life and became a wealthy physician sought by royal and noble persons all over Europe. Just five years later, however, he died under mysterious circumstances at an Austrian inn. He was only forty-eight.

PARADIGM—A paradigm is a system of interlocking facts, theories, and philosophies that is so widely accepted it becomes an implicit framework for thinking about scientific problems. Paradigms are most noticeable when new discoveries call into question an existing paradigm, demanding a new paradigm. The result can be what is called a paradigm shift. An example of a paradigm shift would be the rise of Lyell's geological uniformitarianism and Darwin's evolutionary theory, which together replaced the paradigm of supernatural creation. The rise of quantum theory early in this century created a new paradigm in physics.

See THEORY; HYPOTHESIS; UNIFORMITARIANISM; EVOLUTION; QUANTUM THEORY.

PARAMETER—Here is another one of those scientific words that gets incorporated into popular parlance with only a weak and skewed understanding of what it means. It is most often confused with the word "perimeter." Strictly speaking, a parameter is an element of a mathematical equation whose value can change, and which, when it changes, causes other elements of the equation to change accordingly. More loosely speaking, then, a parameter is anything whose changing value alters related values.

PARSEC—Though often misunderstood as a unit of time (its misuse this way in the movie *Star Wars* has become a classic goof among science-fiction buffs), the parsec is a unit of distance in astronomy. It is equal to 3.26 light years. The term comes from the apparent shift in position of stars when viewed from opposite sides of the Earth's annual orbit of the sun—a phenomenon known as parallax. The parsec is the distance at which an object must be (3.26 light years from Earth) to produce a parallax shift of one second of arc. The term is made up of the first syllables of parallax and second.

PARSIMONY—This is the principle that scientists use when deciding between alternative explanations for a phenomenon. The rule of parsimony is that one should choose the simpler explanation. It is not necessarily always the right one, but in the vast majority of cases it does work out that if some effect can follow a simple chain of events, there will rarely be an opportunity for the more complex chain of events to occur. Also, the greatest advances in scientific theory are usually based on the adoption of simpler sets of events to account for a phenomenon. The principle is also known as Occam's razor, named for William of Occam (c. 1300–1349), an English philosopher.

PARTICLE—In several branches of science this term is used to refer to various bits of matter, but its most interesting usage is in atomic physics, where it is applied to the subunits that make up an atom. In this case a particle is thought of as a very tiny lump of matter such as a proton, neutron, or electron. (It is believed that protons and neutrons are made of still smaller particles called quarks.) In most serious discussions of particle physics, however, it soon becomes clear that particles are not simply lumps of matter. In accordance with Einstein's relativity theory, which asserts that matter and energy are alternative forms of the same thing, atomic particles are sometimes thought of as bundles of energy—a crude description, adapting the English language to something that remains a fundamental mystery.

Not the least of the mystery is that particles do not always behave as it would seem plain particles should. Instead, they sometimes behave like waves, which are not kernels of matter at all, but phenomena that propagate through matter. One prominent exception to this is the light wave, which passes through space devoid of matter. How this can be is a mystery that science has attempted to explain in two ways. Once it was assumed that matter, called ether, permeated all space. When it was shown that ether does not exist, light was thought of as moving particles, photons, that, mysteriously, also have wavelike properties.

See WAVE; MATTER; ATOMIC STRUCTURE; RELATIVITY; ETHER.

PARTICLE ACCELERATOR—A machine, often called an atom smasher in newspapers, that physicists use to find out what atoms are made of. The accelerator is used to shoot a beam of particles such as protons or electrons at target atoms or subatomic particles; the physicists then determine what kind of particles come flying out of the collision. The accelerator makes the particles in the beam move faster and faster along a course until they smash into their target atoms. The more energy the projectile particles have, the more likely they are to break apart tightly bound subatomic particles such as quarks.

To make the projectile particles move faster, accelerators provide an electromagnetic field that attracts or repels them according to their electrical charge. A series of attractive electromagnetic fields will accelerate particles down the line. Some accelerators are arranged in a straight line; others are ring-shaped and use magnets to steer the particles in a circle so that they can reenter the magnetic field repeatedly, receiving another boost of energy each time. Then they are diverted out of the ring and toward the target.

As physicists probe deeper into the atom, seeking to break apart ever more strongly bound subatomic particles, they require stronger and stronger accelerators. As these machines get bigger and stronger, their costs run into the many millions of dollars. Large particle accelerators are among the most expensive tools of science, in a league with spacecraft, although the costs pale by comparison with many items of military hardware.

See ATOMIC STRUCTURE; QUARK; PROTON; ELECTRON.

PASTEUR, LOUIS, 1822–1895, French chemist and microbiologist—Many of Pasteur's achievements grew out of his belief that fermentation and disease were basically the same. In other words, Pasteur said, milk turned sour for the same reasons people got chicken pox. It turned out to be a pretty good idea, leading Pasteur to prove that microorganisms cause fermentation and to speculate that similar microbes attack the body. He developed a method of protecting foodstuffs against microorganisms—today known as pasteurization—and vaccines to prevent several diseases.

Pasteur was the descendant of generations of tanners, and spent most of his childhood showing an interest in nothing but drawing pictures. In school, however, he became interested in science, and earned a doctorate that allowed him to make a living as a chemistry professor in various secondary schools and colleges.

His first important discovery had nothing to do with microbes, but did lay the foundation for a branch of chemistry that is only today being recognized as deeply important in understanding certain aspects of biochemistry. Pasteur was studying tartaric acid, a substance formed during the fermentation of grape juice, and comparing it to racemic acid, a byproduct of certain industrial processes. The puzzle was that the two acids had identical chemical structures but very different physical properties.

When polarized light was shined on crystals of each substance, for example, it passed straight through racemic acid but was rotated to the right in tartaric acid. Pasteur eventually figured out the reason for this: racemic acid was actually a mixture of two forms of the same molecule, forms that were mirror images of one another. They were like the right hand and the left hand. Tartaric acid was identical to one of the two forms of racemic acid. This form had a uniform rotating effect on polarized light. The opposing mixture had light-rotating effects that cancelled each other out. Pasteur then found that a mold would grow on solutions of both chemicals, but that the tiny mold organisms consumed only the tartaric acid. This was also the only form produced by the grape plant. Pasteur had found that although two forms are possible for many molecules, life is built on only one of the two. Today this insight helps explain much of how living cells conduct their metabolism—by a mechanical fit of one specific form of molecule with another. Just as a right hand does not fit into a glove made for the left hand, a right-handed molecule, though just as easy to make chemically as a left-handed one, may have no role in cell metabolism.

In 1863 Pasteur became dean of the new science faculty at the University of Lille, an industrial town. Always sensitive to the needs of industry and industrial workers, he soon started a program that was almost a century ahead of its time—evening classes so that workers could earn college degrees. Pasteur also took day students on regular tours of Lille's factories, and repeatedly stressed the need for scientists to respond to the practical problems of the world.

Among those problems was the spoilage of milk and other foodstuffs. Scientists had long believed that spoilage, or fermentation, was the result of a purely chemical process. Pasteur proved it wasn't. He showed that if a broth were heated first and then sealed off from the outside air, it would not spoil. Open the container, though, and spoilage would soon take place. Pasteur believed that microorganisms floating in the air were

responsible for the spoilage. When they got into some
food, they started eating it, multiplying, and excreting
their waste products. The sum of all these changes was
spoilage. Pasteur developed specific procedures for destroying the microorganisms that caused spoilage in wine
and vinegar.

Next, Pasteur devoted himself to the problem of beer,
devising methods that kept the beer good long enough for
the British government to ship it all the way around Africa to colonial India. Today many products are protected
from spoilage by the heating process that now bears Pasteur's name.

True to his practical philosophy, Pasteur accepted a
government assignment to study a disease epidemic
among the silkworms used in the silk industry in the south
of France. In three years, despite the confusion caused by
two overlapping epidemics and the handicap of a sudden
stroke that partly paralyzed him, Pasteur determined the
mode of transmission of the two diseases and how they
were spread, devised preventive measures against them,
and in the end rescued the French silk industry.

Pasteur's most spectacular achievements, however,
came in the field of immunology—though the field would
not be called that for many decades. He had observed that
animals, once stricken with an infectious disease and recovered from it, remained immune to that disease. But his
real breakthrough came as the result of an accident. He
was working on the problem of chicken cholera, and his
assistant had accidentally left a batch of the causative bacteria standing in a laboratory over a long hot summer.
When injected into chickens, the bacteria produced only a
very mild and transient form of the disease. Pasteur
linked the two observations—that infection conferred immunity and that the mistreated bacteria caused only mild
infection. Perhaps the mild infection, he reasoned, was
enough to confer immunity. When he then gave the same
batch of chickens fresh, fully virulent cholera bacteria,
they were unaffected. Pasteur had inadvertently discovered so-called attenuated, or weakened, bacteria. Ap-

parently, the long hot summer in the laboratory had damaged the bacteria so that they could not cause much of a disease, but had left them still able to trigger immunity.

Pasteur went on to produce comparable vaccines for anthrax, a disease of hoofed animals, and in his most dramatic experiment, produced a vaccine for rabies. Pasteur had been working on the disease and had a collection of attenuated rabies viruses in his lab when word reached him of a nine-year-old boy who had been bitten by a rabid dog. The boy was likely to die, and against the advice of many distinguished colleagues, Pasteur decided that the first human test of his vaccine would be on the child. Since rabies—which affects the brain—does much of its damage before the immune system can mount an attack against it, Pasteur reasoned that his vaccine might provoke the body into attacking the virus sooner. He gave the boy fourteen vaccine injections over a period of weeks, and the youngster lived, becoming a celebrity. The French government immediately established the Pasteur Institute to pursue further work on the disease. It has since become a premier research center.

The boy's name was Joseph Meister, and he became a caretaker at the institute. In 1940, long after Pasteur had died, Meister confronted Nazi troops invading Paris. The soldiers demanded access to Pasteur's tomb at the institute. Meister, wanting to protect the honor of his savior but hopelessly overpowered, committed suicide rather than yield to the invaders.

See GERM THEORY; IMMUNE SYSTEM; ANTIBODY; VIRUS.

PASTEURIZATION—Many bacteria that cause food spoilage can be killed by heating the food to temperatures well below boiling. Louis Pasteur, the French microbiologist, discovered this when developing a way to preserve wines without resorting to boiling, which ruined their flavor. The method is now best known as a means of retarding the spoilage of milk.

See PASTEUR.

PATHOGEN—Any organism or related form that causes disease (pathology). Common examples are certain viruses, bacteria, fungi, and other microorganisms.

See GERM THEORY.

PAVLOV, IVAN, 1849–1936, Russian psychologist— Open a can and the sound leads your cat or dog to behave as if food is coming. Wiggle a finger playfully near a child's ticklish zones and the youngster giggles uncontrollably before you touch him. These are examples of conditioned responses, sometimes called Pavlovian responses after the Russian scientist Ivan Petrovich Pavlov, whose experiments on dogs became the foundation of a broad theory that attempts to explain how people learn.

Pavlov was a Russian doctor who did considerable research on the nature of digestion around the turn of the twentieth century. He won a Nobel Prize in 1904 for this work, but is better known for discoveries in behavior that were an offshoot of his digestion studies. These latter discoveries were admired and adopted in Soviet Russia after the revolution, even though Pavlov was an outspoken opponent of the new regime, perhaps even the first in a long tradition of scientific dissidents in the U.S.S.R.

As Pavlov was doing his experiments on dogs, he happened to notice that they began to salivate not when they tasted food but when they saw somebody bringing a dish of food. To Pavlov, who believed that reflexes (involuntary responses of the body) could be triggered only by physical acts, it was as startling as if a person's lower leg would jerk spontaneously when he saw the doctor coming with the little rubber mallet. Pavlov wondered how a reflex could be set off by a seemingly nonphysical trigger— one that seemed to have its effect not through a physical factor such as taste or the stab of a pin, but through some mental process that anticipated the physical phenomenon.

Pavlov tested this hypothesis with an experiment that has become a classic. He obtained some unconditioned dogs and every time he fed them, he rang a bell. After several days of this, he rang the bell without giving them any food. The dogs started salivating. Pavlov had proven his hypothesis: reflex actions can come under the control of a mind that has been trained, or conditioned, by experiences. The animals' brains had adopted the sound of the bell as a trigger of the salivary reflex.

Then Pavlov tried another experiment. He fed dogs and, at the same time, showed them a circular light. He also showed them an oval light at times when no food was imminent. The dogs quickly learned to salivate when the circular light came on but not when they saw the oval one. The circular light excited the animals' salivary reflex, the oval light did not. Then Pavlov gradually changed the oval light, making it more and more circular. When the dogs could no longer distinguish the two lights, they became visibly upset, howling and pacing around nervously. It was as if the animals' minds had developed a conflict between excitatory and inhibitory responses, which led to erratic behavior.

Pavlov soon extended his ideas to seek explanations for human behavior. How much of human learning was simply a result of conditioning? he wondered. How much of human neurosis was a result of conflicts between countervailing stimuli from the outside world? Although many psychologists today reject Pavlovian conditioning as too simplistic an explanation for all human behavior, many accept that it does play a limited role in such behavior. When children begin to acquire language, for example, it may well be through conditioning. If a parent says "cookie" before handing one to the child, the youngster associates the two and smiles eagerly at the sound of the word.

Pavlov even extended his ideas to the problem of human psychosis. Eventually Russian psychiatric hospitals adopted his notion that mental derangement could arise from a storm of conflicting stimuli reaching the brain, and treated psychotic patients by putting them in quiet, non-stimulating surroundings.

Pavlov's interest in reflexes began early, when, as a student in St. Petersburg (now Leningrad), he read a book on the subject. He had been born in a small town in central Russia, the son of a village priest, but his family soon moved to the city. Although the family at first put him in training for the priesthood, young Pavlov's interest in science, especially biology, turned him toward a medical career. He became a doctor but devoted much of his time to research in his own private laboratory. He focused first on studying the circulation of the blood, and then moved on to study the physiology of digestion, which starts with the secretion of saliva—actually a digestive juice that breaks down starches into sugars.

When the Communist revolution created enormous turmoil in Russia, Pavlov requested permission from Lenin, the new Soviet leader, to move his laboratory abroad. Lenin denied the request, saying that the nation needed him. But since it was a time of famine, Lenin offered Pavlov the larger food rations granted honored members of the Communist Party. Pavlov refused the privileges on egalitarian grounds. It was the first in a series of open rebukes to the new leaders, and even though the government honored him many times, Pavlov repeatedly denounced Communism. "For the kind of social experiment that you are making," Pavlov once declared, "I would not sacrifice a frog's hind legs." In 1924, when the Soviets expelled the priests' sons who were students at the medical academy where he was a distinguished professor, Pavlov resigned, announcing that he too was the son of a priest and would leave with the students. He continued his work in his private laboratory.

There were further gestures of protest but, in the last two years of his life, Pavlov muted his protests. It was the mid-1930s; the government had increased its support for science and, apparently, Pavlov's patriotism was rekindled when war with Japan seemed imminent. Although never a member of the party, Pavlov finally voiced his support for the Communists. Although his name has sometimes been associated with the technique of brainwashing, there is no evidence that Pavlov was responsible for this practice.

PAVLOV'S DOG—The animal in which the Russian scientist Ivan Pavlov first elucidated the nature of the conditioned response.

See PAVLOV.

PENZIAS, ARNO, 1933– , German-born American. See BIG BANG.

PEPTIDE—A compound formed of two or more amino acids linked together by chemical bonds. When a great many amino acids are linked, the compound is called a protein. Pieces of a finished protein are called peptides.

See AMINO ACID; PROTEIN; GENES, HOW THEY WORK.

PERIODIC TABLE OF THE ELEMENTS—Different elements share common properties in their ability to enter into chemical reactions. If you list the elements in ascending order according to how heavy each atom is, these chemical properties recur at periodic intervals along the list. Dimitry I. Mendeleyev, a Russian chemist, was among the first to recognize this, and was the most successful in devising a table in which all the elements could be displayed so as to show the similarities in certain "families" or groups of elements. His periodic table was not only a convenience, it revealed gaps in the list of elements (places where no known element would fit) that correctly predicted the existence of elements that had not yet been discovered.

See MENDELEYEV; ATOM; ELEMENT.

PERRIN, JEAN BAPTISTE, 1870–1942, French physicist. See ATOM THEORY.

PHENOTYPE—The manifested characteristics of an individual, as contrasted with the genotype, or complement of genes of the individual. The genotype and phenotype of an individual may differ because environmental factors prevent full expression of the genes. Also, two individuals may share the same phenotype for a given trait, but be of different genotypes if one is a carrier of the gene for an alternative trait. For example, a brown-eyed person may carry an unexpressed gene for blue eyes, and pass it on to a child.

See GENOTYPE.

PHEROMONE—A chemical substance released into the air by an animal (or into the water if the animal is aquatic) and capable of producing a behavioral response in another animal that smells or inhales the substance. Sex-attractant pheromones are known among many species, especially insects, but not—despite much suggestion to the contrary—among human beings. They may exist, but they have not yet been discovered. Pheromones may also communicate other messages than sexual attraction. Queen bees, for example, produce a substance that inhibits the larval development of potential rival queens.

PHLOGISTON—One of the great theories of early science, later soundly disproven, was the idea that all combustible objects contained a substance called phlogiston. According to this, it was thought that when an object was burned, the fire that emanated was phlogiston escaping from the object. Once this was gone, all that remained of the object was ash. Antoine Lavoisier, the great French chemist, soundly disproved the theory and developed the modern understanding of combustion as the rapid oxidation of substances.

See LAVOISIER.

PHOTON—One way of conceiving of electromagnetic radiation (whether light, radio waves, or x-rays) is as a flow of particles called photons. Sometimes it is helpful to think of a photon as a tiny bundle of energy, which cannot be further subdivided and which therefore represents the smallest unit of electromagnetism. In the parlance of physics, then, a photon is a quantum, or the smallest indivisible unit, of electromagnetic radiation.

Electromagnetic radiation, however, does not always

behave like a flow of particles. It also exhibits properties
of a wave—a disturbance in a medium, such as the ripples
that pass through the water in a pond. The nature of electromagnetic radiation (including the radiation that our
eyes respond to as visible light) is a deep mystery, combining properties of waves and particles.

The thermonuclear reactions inside the sun emit huge
quantities of photons that fly out in all directions, some of
which are obviously visible light. Radio broadcasting tow-
ers emit photons that impinge on antennas to produce an
electrical effect. X-ray machines produce photons with
wavelengths capable of penetrating flesh but not dense
bone. All photons move at the speed of light.

See ELECTROMAGNETISM; QUANTUM THEORY; LIGHT,
NATURE OF; RELATIVITY.

PHOTOSYNTHESIS—This is the process, occurring in
all green plants, that ultimately powers all of life, captur-
ing solar energy and storing it in the form of chemical
bonds in sugar molecules. The plant then binds several
sugar molecules to make starch and stores it. The cells of
the same plant, or of animals that eat the plant, can break
the energy-containing bonds in the starch molecule to release the energy as heat or to transfer it to other molecules. Animals use the released energy to power all their
metabolic reactions, including those that move muscles.
The human body, then, is actually powered by the sun. In
the same sense, the energy released in the burning of
wood is solar energy captured through the process of photosynthesis in the tree's leaves. Even the energy released
in gasoline, oil, coal, and other fossil fuels is solar energy;
it was captured by plants millions of years ago, held in
storage in the form of various substances after the plants
died, and saved in the form of these fuels when the plants
were compressed under mountains of sediment.

Photosynthesis is a two-step process carried out inside
special structures within plant cells called chloroplasts.
Because they contain the green substance known as chlorophyll, chloroplasts are what make green plants green.

When light strikes chlorophyll molecules, they act as catalysts that split water molecules into atoms of oxygen and hydrogen. Most of the liberated oxygen is a waste product, and is dumped outside the leaf to add to the oxygen animals breathe. The hydrogen is combined chemically with molecules of carbon dioxide, CO_2, (a waste product breathed out by animals and absorbed from the air by leaves) to make the sugar known as glucose, whose chemical formula is $C_6H_{12}O_6$. For every molecule of glucose it makes, the plant must take in 6 molecules of carbon dioxide and 6 molecules of water. It will also give off, as a waste product, 6 molecules of oxygen, O_2.

See CHEMICAL BONDING; CONSERVATION OF ENERGY.

PLANCK, MAX, 1858–1947, German physicist—Among the most pivotal scientists in history, Planck was the originator of quantum theory, one of the two greatest achievements of twentieth-century science. This is the understanding that energy is not a continuously variable entity, but that it instead comes in tiny, indivisible packets called quanta. Since Planck advanced it in 1900, quantum theory has led to the most fundamental understanding of what happens inside atoms.

Planck was born in Kiel, in what is now West Germany. He attended the University of Munich and came into contact with the rich, nineteenth-century German cultural life. He developed an inclination to music and almost became a musician. Instead, he chose theoretical physics, then a relatively new field.

After discovering the quantum nature of energy and developing a formula for relating the energy content of a quantum to the frequency of the corresponding electromagnetic wave (through use of a value Planck discovered and now called the Planck constant), Planck played little further role in the growth of quantum theory. Einstein, Bohr, and others were to take over this field of work. But while others went ahead assuming that quantum theory revealed the true nature of energy, Planck maintained that it was only a mathematical convenience that did not necessarily correspond exactly with the real world. Planck did, however, receive the 1918 Nobel Prize in physics for his achievement.

In his later years, Planck's life was not a happy one. Of his seven children, six had died. Then came the Nazi regime. Although many German scientists had emigrated, Planck refused to go. He considered it a duty to remain behind and oppose the Nazis. Over the ensuing years he was repeatedly asked to sign a loyalty oath to the Nazi party, but he always refused. In 1944, when Planck was eighty-six, the Nazis came again to try to get his signature. This time they had a hostage—Planck's own son, who had been arrested for conspiring to assassinate Hitler. If Planck signed, his son would be released. Proud and resistant to the end, he refused. Planck's last surviving child, Erwin, was executed.

After the war, Germany sought to honor Planck and planned a massive celebration for his ninetieth birthday, but the old man died just months before the event. In tribute, Germany's academy of science, formerly named for Kaiser Wilhelm, was renamed the Max Planck Academy, and the nation's highest scientific award was named the Max Planck medal.

See QUANTUM THEORY; ELECTROMAGNETISM; ATOMIC STRUCTURE; BOHR; EINSTEIN.

PLANET—Any of the non-luminous bodies orbiting the sun (and presumably, bodies orbiting many other stars, though no planets are known for certain beyond our solar system). Conventionally, the term is applied only to the nine largest bodies orbiting the sun, although the asteroid belt contains rocky objects not much smaller than planets, and although some planets are about the size of the largest moons, or satellites, of other planets.

See SOLAR SYSTEM, ORIGIN OF.

PLASMA (ATOMIC)—Under certain conditions, such as exist inside stars, atoms disintegrate into nuclei and liberated electrons. Plasma is simply a state of matter consisting of free atomic nuclei and electrons. Tiny regions of plasma are created briefly inside electrical sparks, including lightning. As soon as the energy of the spark is spent, the nuclei recapture electrons to become proper atoms again.

See ATOMIC STRUCTURE.

PLASMA (BLOOD)—A clear, almost colorless fluid consisting of blood from which the red and white blood cells have been removed. Plasma contains platelets and cells called lymphocytes. The portion of plasma that diffuses out of blood vessels into open spaces of the body is called lymph. Lymph is reabsorbed by a system of lymph vessels and carried back to two points just behind the collarbones where the fluid is emptied back into the blood stream.

PLATE TECTONICS—Here is a theory for which almost anyone can examine some of the best evidence. Take a map of the Earth and cut out the continents and larger islands. Next, pretend you have a jigsaw puzzle and try to fit the pieces together into a single land mass. You will find that it is not hard. The east coasts of North and South America fit rather nicely against the west coasts of Europe and Africa. There is even space for Greenland and the British Isles. Madagascar tucks quite neatly against the east coast of Africa.

The fit between both sides of the Atlantic was recognized centuries ago, almost as soon as early explorers produced decent maps. There was even speculation in the sixteenth century that the continents bordering the Atlantic might once have been a single land mass that somehow broke up.

This speculation was soon quashed, however, by the even more obvious fact that the Earth's surface is very rigid—hardly the sort of stuff that could break so monumentally—and that the land masses were very far apart. After all, how could they possibly move so far?

Thus, it was not until the 1960s that new evidence confirmed that there had once indeed been a single great con-

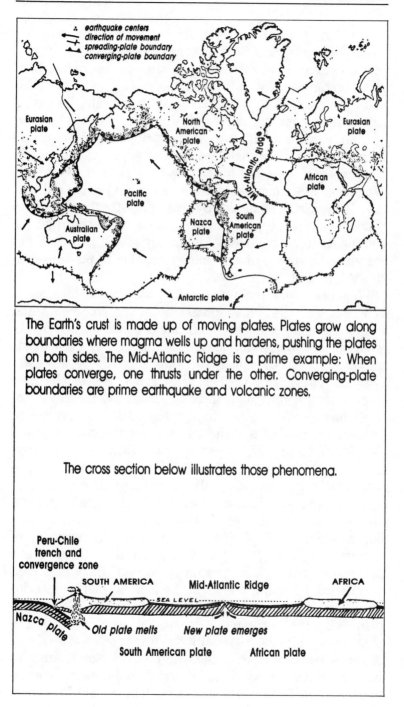

The Earth's crust is made up of moving plates. Plates grow along boundaries where magma wells up and hardens, pushing the plates on both sides. The Mid-Atlantic Ridge is a prime example: When plates converge, one thrusts under the other. Converging-plate boundaries are prime earthquake and volcanic zones.

The cross section below illustrates those phenomena.

tinent that split into pieces which have since drifted apart. Geologists today almost universally accept the theory of continental drift or, as the more complete theory is called, plate tectonics. Not only does the theory explain major features of the Earth's surface, it also provides an explanation for major climactic shifts in the planet's history, and for certain oddities in the distribution of fossil and living plants and animals around the world.

Acceptance was hardly the case in 1915 when Alfred Wegener, a German geologist, tried to revive the old theory. Physicists objected by claiming that earthquake waves showed the planet's crust to be much too rigid for continents to have moved. And in any event, nobody could imagine the source of the energy capable of pushing continents around.

Wegener's four main arguments were soon dismissed. His first was the jigsaw-puzzle fit of the continents and larger islands. He even found that the modern continents would fit still better if one cut them not along their existing coastlines but along the edges of the continental shelves. These are, in fact, the true edges of the continents, and simply happen to be underwater because today's oceans are high enough to have flooded over them. Wegener called his original supercontinent Pangaea. Wegener's second argument was that measurements suggested that the distance between Greenland and Europe was widening very slowly. His third argument was more of an observation: That most of the Earth's crust exists on two levels—the continents, which are high, and the ocean floors, which are low. The continents are largely made of granite, which is lighter than the basalt of which the ocean floors are made. Wegener suggested that the lighter rock was actually floating higher in the Earth's molten interior than were the heavier basalt ocean floors. His last argument was the correspondence to be seen between the plants and animals (especially fossil forms) on either side of the Atlantic. For example, the trees of North America were much like those of Europe, and would form a continuous habitat if the continents were joined, said Wegener.

Similarly, certain fossil animals in South America were most closely related not to others on the same continent but to those in Africa. Wegener believed that these similar groups had evolved as one and then become separated when the continents split.

Still, Wegener's ideas soon fell into obscurity, not to be revived for almost half a century. The revival came slowly, with the discovery of new evidence for continental drift. Explorations of the Atlantic floor revealed a long ridge running roughly north and south, exactly halfway between the opposite coasts of that ocean. The ridge even paralleled the coasts. What was more, the ridge followed a long crack in the ocean floor where molten rock was welling up from below. Rock samples from near the ridge were newly formed, but farther away from the ridge the samples got progressively older all the way to each coast.

The only possible explanation for all this was that the Atlantic floor was slowly expanding on either side of the ridge. At some time in the remote past (geologic evidence suggests that it was around 130 million years ago—or late in the age of dinosaurs), there must have been no Atlantic ocean at all, but simply a single continent of granite, through which a sinuous line of volcanos began erupting. When the heavier, basalt part of the continent (solidified magma) grew large enough, it slowly sank. Water from the primordial ocean, of which the Pacific is today a remnant, then flowed in and the Atlantic was born as a narrow waterway separating the two masses of granite. Recent measurements have confirmed that North America and Europe are moving apart at the rate of one or two centimeters a year.

There was other evidence, too, for continental drift. Basalt contains tiny grains of iron oxide, which is very weakly magnetized by the Earth's magnetic field. As lava containing the basalt cools, the iron particles orient themselves parallel to the north-south lines of the Earth's magnetic field. Once the volcanic rock has solidified, however, the magnetic polarity of the iron particles is frozen into position. As it happens, the Earth's magnetic field has reversed its polarity several times over the distant past, at intervals of hundreds of thousands of years. Scientists who took samples of basalt from many points around Pacific faults much like the mid-Atlantic ridge found that the newest basalt had been magnetized with the Earth's current polarity. Farther away they found a parallel band of basalt with the reverse polarity. Still farther away the polarity was again that of the modern Earth. These stripes of alternating polarity, all parallel to the fault, are another evidence of spreading of the ocean floor and the creation of new crust.

Global maps of the mid-ocean ridges and major fault systems are now believed to show the edges of the seven major and several more minor pieces of the Earth's crust—consisting of either granite or basalt—which are called plates. Most of North America is on one plate. Europe is on another. The two are being pushed apart by the growth of the Atlantic plate between them. The western edge of the North American plate is the San Andreas fault in California. The western side of that fault is actually the eastern edge of the Pacific plate—or of a sliver of a plate between the Pacific plate and the North American plate—uplifted because it is buckling under pressure as the North American plate pushes against it. Geologically speaking, San Francisco, Los Angeles, and most of the California coast are not part of North America. The movement of the two plates, which are rubbing together, causes California's frequent earthquakes.

Since the Atlantic is widening and pushing its bordering continents apart, the Pacific is shrinking. Where is the Pacific plate going? Along most of the western edge of the Pacific it is colliding with and plunging under the edge of the Asian plate. The titanic collision is responsible for the many volcanos that ring the northern and western edges of the Pacific Ocean. The plunging basalt of the Pacific plate is re-melted and added back to the magma inside the Earth.

When two granite plates collide, neither can move under the other. They buckle, pushing up huge mountains.

India, for example, was a small continental plate, formerly attached elsewhere to Pangaea, which broke off, drifted northward, and rammed into the southern edge of Asia, thrusting up the Himalaya mountains.

All over the Earth, then, crustal plates are on the move—pushing, grinding against one another and causing earthquakes, buckling up to form huge mountains, plunging under other plates to recycle basalt, and cracking open holes through which new fountains of magma form volcanos. Plate tectonics also helps explain why Earth's climate in the past was so different from what it is today. It is known that the flow of warm or cold ocean currents affects climate, and it follows that these flows can be blocked by land masses moving into their path, or can be redirected into new regions as oceans open up. Plate-tectonic theory is therefore what scientists call a powerful theory: it explains and intellectually unifies many phenomena that once seemed disparate and unrelated.

But the question still remains: What makes the plates move? The best explanation at the moment suggests that there are mighty currents flowing in the molten rock below the Earth's crust. They are called convection currents, and are similar to currents that can be seen in the atmosphere and in bodies of water. Convection currents result from changes in the density of a fluid. Warmer water, for example, rises because it is less dense (and hence slightly lighter for a given volume) than cooler water. But when it nears the surface it cools, becoming more dense, and sinks again. It can't sink straight down because more warm water is rising up beneath it, so it flows across the surface a short distance and sinks elsewhere. Gradually, distinct patterns of currents form in the water. You can even see convection currents in a cup of coffee or a bowl of soup. A cup of Japanese miso soup gives one of the best demonstrations of such currents, because the bean particles flow with the currents, making them visible. The Earth's atmosphere behaves in the same way, with warm equatorial air rising and flowing toward the poles to replace the sinking cold polar air, which then flows toward the equator.

Geologists suspect that the same things are happening in the molten rock inside the Earth. As the rising magma meets the cool crust it is deflected sideways, cools, and sinks again. The crustal plates, which float on the magma, are carried along like massive rafts. Magma doesn't flow as fast as water, however. It is much denser and more viscous, and may flow at the rate of only an inch or so a year.

See WEGENER.

POLYMER—One of the most common things that happens to molecules is that they attach themselves to other molecules of the same kind. When these attachments take the form of long chains of the same molecule—resembling beads on a string or links in a chain—the total molecule is called a polymer. A short chain of two molecules is a dimer. Three linked molecules is a trimer. More than three is a polymer. Sometimes, when chemists need to be specific, they will speak of a 14-mer or a 26-mer. Examples of polymers include proteins (which are chains of amino acids) and plastics. Polyethylene, a common plastic, is a polymer made up of ethylene subunits.

PROTEIN—Protein molecules are the molecular workhorses of life. The genetic messages carried in DNA do nothing but specify which kinds of proteins will be synthesized in a cell, and when and how much will be synthesized. The function of the cell, whether it is to be an amoeba, a brain cell, or a cell in an apple-tree leaf, is determined by the mix of specific kinds of protein molecules within it. Some proteins link themselves with fat molecules to form structures such as cell membranes. Others, such as hemoglobin in the blood, incorporate iron atoms to help carry oxygen from the lungs to the rest of the body. Still other proteins act as enzymes whose job is to digest the proteins we eat, splitting beef protein, for example, into the amino acid molecules of which it is composed.

PROTEINS are made, according to instructions of the genetic code, as a linear sequence of subunits called amino acids. These link like boxcars but then fold up into various forms, regular and irregular, including spirals and sheets.

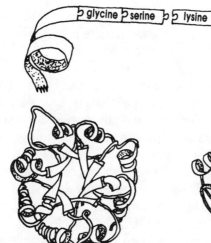

Triose phosphate isomerase (helps break down sugar in muscles)

Two black balls in this calcium-binding protein from a carp represent captured calcium atoms.

Another way of depicting proteins shows atoms as balls. This is a gene-regulating protein. Its shape fits perfectly into the grooves on a DNA double helix, blocking its message from being read.

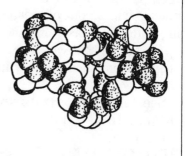

These loose amino acids are then available to be reassembled within the various cells of the body into other kinds of proteins. There are more than 30,000 kinds of proteins in the human body.

See GENES, HOW THEY WORK; AMINO ACID.

PROTON—One of the three particles that make up an atom. A hydrogen atom consists of one proton (the nucleus) with one electron orbiting around it. All other atoms have nuclei consisting of several protons and approximately the same number of neutrons. Each proton carries a positive electrical charge that is precisely balanced by an orbiting electron, which carries a negative charge of exactly the same strength. Each proton is made of three quarks—two "up" quarks and one "down" quark. (Up and down are whimsical terms for opposite attributes that are otherwise much more complex to describe.) Protons are much heavier than electrons. Like neutrons, they have a mass that is 1,836 times that of an electron.

Although protons have long been considered stable particles—incapable of disintegrating, or decaying—there is growing suspicion that they are mortal, though very long-lived. Some calculations suggest that any given proton has a 50 percent chance of disintegrating in 100-billion-trillion-trillion years. (The universe itself is less than 20 billion years old.) Or, considered in another way, in one year, one proton should decay in a population of 100-billion-trillion-trillion protons. What this means, if the theory is correct, is that some day very far in the cosmic future, matter will go out of existence. Some physicists are seeking to detect a proton in the process of decaying by setting up detectors that can record the characteristic emission of energy from such a decay in a huge vat of highly purified water.

See ATOMIC STRUCTURE; NEUTRON; ELECTRON; QUARK; HALF-LIFE.

PROTOPLASM—This is a somewhat antique term that refers to the entire contents of a living cell. When the nucleus is set apart, the rest of the cell contents are often called the cytoplasm. As biologists studied the cell in ever greater detail, it became clear that it contained not some generalized jelly but many specific structures, each with a job to do. In addition to the nucleus, which holds the genes, these structures include mitochondria, which extract energy from nutrients, and ribosomes, which take the genetic instructions encoded in DNA and synthesize proteins according to them.

See MITOCHONDRIAN; NUCLEUS (LIVING); RIBOSOME.

PTOLEMY, 2nd century A.D., Greco-Egyptian astronomer. See COPERNICUS.

PULSAR—If you aim a highly sensitive radio antenna out into space, you are likely to pick up radio signals that pulse regularly at rates varying from once every few seconds to fifty times per second. A loudspeaker attached to the radio receiver would produce bursts of static (a jumble of radio noises) at these regular intervals. This is essentially what some very startled British radio astronomers found a few years ago. The pulses were so precisely regular that at first the astronomers allowed themselves the fantasy that they were being produced by an extraterrestrial intelligence. The signal sources were even named "L.G.M.," for little green men.

Eventually it became clear that the pulses were coming from very small, starlike objects that emitted radio signals, as do all stars, but which were spinning very rapidly. The magnetic field of each such object organized the emanation of radio signals into a strongly directional pattern. Every time the object rotated on its axis, an apparently directional beam of radio signals would sweep the heavens like a lighthouse beacon.

Discarding L.G.M. for a more scientifically sober name, the astronomers called the objects pulsars. Further theoretical work has led to general agreement that a pulsar is what is left of a fairly small star after it has exploded into a supernova—a stage that all stars reach when they exhaust their nuclear fuel—and then collapsed into a cooler, paler shadow of its former self. Because much of the matter remaining after this collapse consists of loose neutrons, pulsars are also called neutron stars. When a larger star reaches this stage, it may have so much mass remaining that it collapses into a black hole, with gravity too powerful to let light or radio signals out.

See SUPERNOVA; BLACK HOLE.

PUNCTUATED EQUILIBRIA—A relatively new name for an older version of the theory of evolution, suggesting that most new species arise through a short burst of evolutionary change and then remain relatively unchanged for the rest of their existence. In other words, most species, at any given moment, are not in the process of evolving into new species. They remain fundamentally unchanged from their ancestors of long ago—sometimes even of millions of years ago. Within a given lineage bursts of evolution may occur once every few million years. The theory of punctuated equilibria is a rival to the theory that most or all new species develop through gradual, continuously operating processes.

Like most good theories, Darwin's theory of evolution has left room for people to continue arguing its fine points. One of the points that Darwin glossed over was the precise mechanism by which a new species arises. His theory of natural selection suggested that small changes are continually being accumulated in various populations of plants and animals, and that eventually, populations that once were alike become different, and so much so that they no longer interbreed, but instead henceforth constitute separate species. Fossil evidence of this gradual, continuous change was lacking in Darwin's day, and even the great man himself recognized this. He suggested, though, that this was merely the result of incomplete discovery. Further finds would fill in the gaps, said Darwin. But a century of additional fossil-hunting, while it has filled in some gaps, has left most still open. Instead, the pattern of punctuated equilibria has emerged from the fossil record, with most species seeming to appear suddenly and then undergo almost no change for millions of years afterward. They may become extinct just as suddenly, or a related species may suddenly appear with no surviving trace of a transitional form.

Some evolutionists—principally Stephen Jay Gould and Niles Eldredge, both American paleobiologists—struck by the long periods of evolutionary stability, or stasis in the fossil record, now think that new species must be born within comparatively short periods of rapid evolutionary change—so short that the likelihood of ever finding the fossils of intermediate forms is very small. Indeed, the conditions of fossilization are such that a span of 10,000 or even 100,000 years could easily disappear without a single fossil being preserved. Although the theory of punctuated equilibria is sometimes misunderstood as suggesting the instantaneous appearance of new species, the scientists who support it advocate no such thing. Instead, they see the changes taking a number of generations and stretching out over many thousands of years.

Exactly how these changes occur is still a matter of great debate. Some people think there may be certain mutations that have a broad effect on many other genes, leading to major changes at a stroke. Other authorities believe that old-fashioned natural selection can operate much more rapidly than usual when environmental conditions are changing rapidly. Many habitats undergo substantial ecological upheavals from time to time. Perhaps the most widespread upheavals have come after mass extinctions. For reasons yet unknown, there have been relatively brief occasions when large numbers of species have become extinct. As a result, large suites of ecological niches have opened up at once, providing opportunities for mutants of existing species to find a favorable habitat.

By no means are all evolutionary scientists convinced that most evolution took place in such brief and rapid bursts. Some feel that the absence of intermediate fossils represents not a pattern of evolution but simply a failure to find the transitional forms. These objectors start with the generally accepted idea that new species arise in small, isolated populations that only later become successful enough to spread into larger areas. As a result, they maintain, it is highly unlikely that a fossil dig will happen to take place within the isolated, small area where evolution may have been proceeding gradually. Only when the species attains its more viable complement of new features will it spread, leaving potential fossils over a much larger area.

The debate between advocates of gradualism and those of punctuated equilibria is not likely to be settled soon.

See EVOLUTION, THEORY OF; NATURAL SELECTION; MODERN SYNTHESIS.

QUANTUM THEORY, QUANTUM MECHANICS—

In a sense, quantum theory is to energy what atom theory is to matter. Both were revolutionary ideas asserting that each entity existed in discrete quantities, in particles or packets. Atoms, it later turned out, could be subdivided into protons, neutrons, and electrons. But packets of energy, called quanta (the plural of quantum), cannot be divided. Quantum theory contradicts ordinary experience, which suggests that energy—whether as heat, light, electricity, or in any other form—is a continuously variable thing, which you can get in any amount simply by "turning the dial." The truth, instead, is that the energy situation is rather like that with matter. It may look as if you can fill a glass with any amount of water whatsoever, whereas in fact you can fill it only in a stepwise manner, with a definite number of molecules of water in each step. Likewise, energy comes only in the fundamental, indivisible units known as quanta. (A quantum of light, incidentally, is also called a photon.)

Quantum theory, generally considered one of the twentieth century's greatest intellectual advances, was born in 1900 when Max Planck, a German physicist, tried to make sense of experiments in which he was analyzing the energy in the light emitted by heated objects. The data made no sense under the generally accepted theory of

light, which held it to be a continuous entity, disturbance
in a medium, or a wave whose height could vary continuously from low to high. The data Planck was getting didn't
fit this concept. To Planck it looked as if the light energy
existed in discrete particles. To explain his finding Planck
was forced, as he later described it, to "an act of sheer
desperation."

Planck decided that the old theory was wrong, and that
light somehow came in discrete bundles, very small packets that he named "quanta," from the Latin for "how
much." The same root, of course, supplied our word
"quantity." Although quantum has taken on a colloquial
and entirely nonscientific meaning suggesting a very large
amount—as in "a quantum leap"—the real quantum is
small indeed. Quanta come in different sizes, depending on
the wavelength of the light that contains them, but it is
close enough to say that it would take billions upon billions
of quanta to supply the energy needed to lift a grain of
sand one centimeter. Nevertheless, although quanta are
small, there are lots of them.

At first most physicists rejected Planck's heresy. General acceptance did not come until 1913 when Niels Bohr,
the Danish physicist, used Planck's equations describing
quantum theory to predict the energy of the light emitted
by heated hydrogen atoms. After this Planck had little
more to do with the idea he had begun other than to receive the Nobel Prize for it in 1918. Others were to take
the concept much further, using it to explain not only the
nature of light and other forms of electromagnetic radiation but the role of energy inside the atom. Since Einstein's relativity theories show that matter and energy are
interchangeable, relativity and quantum theory have become intimately mingled to form a comprehensive basis
for probing the fundamental nature of all existence.

One of the physicists who recognized Planck's insight
was Einstein who, at the time in 1905, was still a lowly
patent clerk and very much removed from the scientific
community as he vigorously pursued his own radical ideas
about relativity. In that year, however, Einstein showed

that not only was light emitted in quanta, as Planck had said, but that it was absorbed in quanta too. This insight came during Einstein's effort to explain the photoelectric effect—the natural phenomenon that causes metals and certain other substances to emit electrons (the subatomic particles that orbit an atom's nucleus) when light shines on them. It was assumed that electrons were only loosely bound to metal atoms, and that as light shined on them, they absorbed energy until they had collected enough to jump out of the metal. But experiments didn't give results that fit this idea. The emission of electrons depended on the frequency of the light as well as its intensity. If the frequency was too low, it didn't matter how intense the light was, no electrons would jump out of the metal. What Einstein surmised (correctly, it turned out) was that an electron couldn't store up the energy it was getting from the light being beamed upon it. It had to get enough energy to break free of its atom from one, single collision with a photon. Otherwise, it couldn't absorb any energy at all. It was for this explanation—and not the theory of relativity, which he also published in 1905—that Einstein won the Nobel Prize in 1921.

The next great advance in quantum theory came in 1913, with Bohr's work on the light emitted by hydrogen atoms that are heated. When Bohr found that quantum theory predicted the wavelength of this emitted light, he realized that it also said something else about events inside atoms, and in particular about the way electrons behave. The insight led Bohr to modify existing ideas about the structure of atoms, ideas largely put forth by Ernest Rutherford, the New Zealand physicist. Bohr's modifications helped explain how an atom can absorb outside energy, store it for a time, and then radiate the energy away again.

Rutherford's model of the structure of the atom had suggested that electrons orbit their respective atomic nuclei in much the same way that planets orbit the sun. But his theory posed a paradox. If electrons really behaved like planets, then their motion around a curved path would continuously drain energy out of the electron. (The Earth, for example, is slowly spiraling closer to the sun as it orbits around it, and a car rounding a curve tends to slow down.) This continuous loss of energy should quickly cause all electrons to spiral inward and collide with their nuclei. Furthermore, calculations showed that if Rutherford's atomic model were right, this should happen very quickly, and that there should be no atoms as we know them. Obviously something was wrong. Bohr extended quantum theory to find the error and explain how electrons really behave.

Electrons, he said, are not like planets. In the world inside the atom things don't happen the way they do elsewhere. Since quantum theory shows that energy comes only in definite quantities, an electron cannot continuously lose energy as it flies along. It is locked in its orbital path (really a cloudlike shell) until it can give up an entire quantum of energy. Then it will simply drop to a lower shell and stay there until it is somehow deprived of another quantum. By the same token, Bohr said, an electron can absorb a quantum of energy and jump up to the next higher shell. Rutherford's atom, modified by Bohr's use of quantum theory, is the one physicists know today.

When an atom absorbs energy—such as when it is heated, for example,—its electrons jump to higher shells. But because they are unstable in those shells, they also drop back down, emitting their briefly held quanta as photons. When metals become hot enough, the photons their electrons emit contain enough energy to be visible to the human eye as red light. The heated object is said to be red hot. If even higher levels of energy are pumped into the object, the energy of the emitted photons increases and the color of the light changes to yellow and, finally, to white. The filament in an electric light bulb is made of a substance (mainly tungsten metal) that becomes white hot when household electricity (itself a barrage of loose electrons) is passed through it. At a furious pace electrons in the tungsten atoms are kicked into higher shells only to drop back almost instantly. In less than a second, the tungsten heats up to around 5,000 degrees Fahrenheit. At that temperature the electrons in the tungsten are being kicked up and falling down with such energy that the flood of photons coming out of the metal registers as white light with a tinge of yellow and red.

In the early part of this century a number of physicists and mathematicians made use of quantum theory to develop a deeper understanding of matter called quantum mechanics. Based on quantum theory, quantum mechanics seeks to explain conventional Newtonian mechanics as a consequence of the decidedly non-Newtonian phenomena that occur inside atoms. As it turned out, all of the properties of electrons could be expressed only in quantum terms. The main shell in which an electron moved was, as we have seen, defined by quantum conditions. So was the specific subshell in which the electron orbited within the main shell. Physicists also found that electrons move in response to external magnetic fields only in certain discrete ways. And finally, electrons could exist in only one of two spin states, in which, by anology with a top, they are conceived as spinning either clockwise or counterclockwise.

Wolfgang Pauli, an Austrian physicist, found in 1925 that within any atom no two electrons may have the same four quantum states. It is as if atoms are hotels with only so many single rooms for electrons. This discovery is known as the Pauli exclusion principle.

Another famous concept that emerged from quantum research was the uncertainty principle enunciated in 1927 by Werner Heisenberg, a German physicist. It is impossible, Heisenberg said, to determine both the position of an electron and its momentum at the same time. You can determine only one with complete precision and, if you wish to know the other at the same instant, you must settle for a probability statement. Heisenberg said this not because the instruments for measuring the location or momentum of an electron were poor, but, he showed, because there is an intrinsic property of subatomic matter that causes either condition—be it position or momentum—to be disturbed by the act of observing the other. The idea greatly disturbed Einstein because it meant that reality in its most fundamental form was not completely knowable.

Einstein's objection to Heisenberg's principle was largely philosophical, but while the uncertainty principle has today become almost a household phrase (especially among those who haven't the faintest idea what it's really about), philosophers are still troubled by it. What the uncertainty principle reveals is that the very act of measuring one property of an object blurs out its other properties. They lose some of their reality. Though it seems hard to believe, philosophers who understand the physics and mathematics of quantum theory suggest that, true as it seems to be, it has led us close to a realization that in some sense the fundamental pieces of which the world is made exist only in the form in which we happen to try to perceive them at any given time.

See ATOMIC STRUCTURE; ELECTRON; LIGHT; ENERGY; EINSTEIN; PLANCK; RUTHERFORD; BOHR; PAULI; HEISENBERG.

QUARK—These oddly named objects are the most basic building blocks of matter yet discovered. There are six kinds (not counting antiquarks, which are quarks made of antimatter), and they can be put together in various combinations to make a proton or a neutron, the two particles that make up the nucleus of an atom. Quarks can also be put together to make any of scores of other particles that physicists have produced in particle accelerators, but which decay almost instantly and are not ordinary parts of the material world. Electrons (which are themselves fundamental particles and not made from quarks) in orbit around the quark-based nucleus make the whole assembly an atom.

The particle physicists who study these matters, as if to lighten their heavy discourse, select whimsical names for the objects they study. The term "quark," for example, is a nonsense word taken from James Joyce's *Finnegan's Wake*, which contains the line, "Three quarks for muster mark." They also favor whimsical names for the different kinds of quarks. There is the "up quark" and its counterpart the "down quark," for example. These are the only two quarks that make up ordinary protons and neutrons. But if you smash atoms together to produce odd, transient particles, you also find evidence for the "strange quark" and its partner the "charmed quark." Discovered in the same way fairly recently are the "bottom quark" (sometimes called "beauty") which goes with a "top quark" (also known as "truth"). These quarks do not exist as such in ordinary matter. They come into being only briefly as components of the very short-lived particles that are formed in atom smashers but which turn back into energy in a fraction of a second.

Nobody knows for sure whether quarks are really the most fundamental particles that exist. Formerly, protons and neutrons held this status, but it was found that they were each made of three quarks that were somehow stuck together.

See ATOMIC STRUCTURE; PARTICLE ACCELERATOR (the device physicists use to study quarks); PROTON; NEUTRON.

QUASAR—Discovered a little more than twenty years ago, these are mysterious objects in the sky, which most astronomers believe are the most distant, brightest, and oldest objects in the universe. The name quasar is a contraction of the original name, quasi-stellar, given because these objects look like stars when viewed through an ordinary telescope, but have odd characteristics that indicate they cannot be ordinary stars. For one thing, the light coming from them has the greatest degree of red-shift known. Red-shift is an astronomers' term for the shifting—toward the red end of the spectrum—of the light coming from an object. It is a measure of how fast the object is moving away from Earth. (Red-shifts are a consequence of the so-called Doppler effect, which also occurs with sound. When a source of sound or light is moving toward an observer, its frequency is perceived as being higher than it actually is. When the source is moving away, the frequency seems lower than it is. This is because the source is either partially catching up with or moving away from the signals it emits.) Because of the way in which the universe is expanding, the farther away an object is from the Earth, the faster it is receding from Earth. The extreme red-shift of quasars implies they are the farthest known objects from Earth.

Since quasars are the farthest objects known, the light we see from them has taken the longest of any light we see in the universe to reach Earth—billions of years. Hence, when we look at a quasar, we are seeing the elements of the universe as it existed shortly after it began. The fact that the light from quasars is powerful enough to reach us over such distances at such brightness means that they must be producing light at phenomenally high rates. A single quasar puts out as much light as a thousand ordinary galaxies, each consisting of a hundred billion stars. In fact, if all these interpretations are correct, quasars produce energy 100 trillion times more powerful than an ordinary star. Astrophysicists have absolutely no idea how such prodigious amounts of energy could be produced. This mystery has led a few astronomers to doubt the validity of using red-shifts as a measure of speed or distance. They suggest that quasars are actually much closer and less powerful and, consequently, that their great red-shift is a consequence of some other factor.

SEE BIG BANG; DOPPLER EFFECT.

RADIATION—This is a vague term that includes several different kinds of particles (or waves) emitted by several kinds of sources. (Radiation may be described as consisting of either particles or waves, since all forms of radiation have such a dual nature.) The most common form of radiation is the flood of electromagnetic waves (or particles called photons) from the sun, some of it visible as light. When unstable (radioactive) atoms or subatomic particles disintegrate, they also emit radiation or particles of various kinds. The best known of these are labeled with letters from the Greek alphabet. Alpha particles, each of which is equivalent to the nucleus of a helium atom (two neutrons and two protons), are the least penetrating particles, being stopped by seven centimeters of air. Beta particles, or electrons (emitted when a neutron in an atomic nucleus loses an electron and becomes a proton) can penetrate about twenty-five feet of air or be stopped at their source by metal foil. Gamma radiation, usually emitted along with alpha and beta particles, consists of photons of electromagnetic radiation. It has much greater energy than alpha or beta particles and can penetrate six inches of lead.

All forms of radiation can damage living cells, although gamma radiation is far more likely to do so because of its greater ability to penetrate. As gamma radiation travels, it wreaks biochemical havoc within cells. It breaks apart various organic molecules, causes others to bond to one another, and alters the molecular structure of DNA, randomly changing the genetic instructions encoded at any given point in this master molecule of life. Most often this last effect does no more than kill one cell, but if the alteration happens to change part of the DNA message into another, undesirable part, the cell may lose control of its rate of proliferation. The result is cancer. The sun, of course, is highly radioactive, and can cause this effect in skin cells through its ultraviolet radiation.

Apart from solar radiation, all life on Earth proceeds in an environment that always contains varying amounts of natural radiation. Most rocks contain a few radioactive atoms, and people who live in stone houses actually get a measurably higher radiation dose than those who live in wood houses. Still, the amount appears to cause no harm, both because its chance of producing a deleterious mutation is low and because living cells are capable of repairing the damage done by low levels of radiation.

See ATOMIC STRUCTURE; ELECTRON; PHOTON; GENES, HOW THEY WORK; MUTATION.

RADIOACTIVITY—This is a property exhibited by many kinds of atoms that are likely to undergo the spontaneous change in which the atomic nucleus spews out one or more subatomic particles (neutrons or protons, for example) accompanied by a burst of energy. There are several naturally radioactive substances, such as uranium and radium, but many other substances can be made radioactive by bombarding them with the particles emitted by radioactive atoms. The most common forms of atomic radiation are alpha particles or rays (consisting of two neutrons and two protons bound to one another, or the same thing as constitutes the nucleus of a helium atom), which can penetrate only a few inches of air or be stopped by a sheet of paper; beta particles or rays (electrons moving at nearly the speed of light), which can penetrate up to a millimeter of lead; and gamma rays (a burst of photons, or very short-wave electromagnetic radiation), which are the deadliest of all radioactive emissions because they can penetrate up to seven inches of lead.

See DECAY (RADIOACTIVE); RUTHERFORD; FERMI; RADIATION.

RADIOCARBON DATING—A method of estimating the age of any organic material by measuring its content of radioactive carbon (C^{14}), which is normally produced in the atmosphere when cosmic rays hit nitrogen atoms (N^{14}). All living organisms take in C^{14} in a known ratio to ordinary carbon (C^{12}), but after they die, the amount present in them begins to decay back to nitrogen at a known rate. (Half the amount of C^{14} present decays every 5,730 years, plus or minus 40 years.) By checking the remaining ratio of C^{14} to C^{12}, it is possible to determine how long the decay has been going on. The method originally assumed a constant amount of C^{14} in the atmosphere during the past history of the Earth, but it is now known that the amount has changed, with the result that the radiocarbon dating method underestimates the age of objects. A correction factor must be applied to it. Radiocarbon dating is useful for objects up to about 80,000 years of age. Older objects will have no measurable C^{14} remaining. Other forms of radiometric dating can give the ages of objects millions and billions of years old.

See RADIOMETRIC DATING.

RADIOMETRIC DATING—When a scientist talks about fossils or rocks being "dated" at so many thousands or millions or billions of years old, the method that supplies the ages of these objects is often a form of radiometric dating. The "radio" refers to radioactivity. There are many other methods of dating but this is considered among the most reliable, if properly done. Radiocarbon, or C^{14} dating, is one well-known example of radiometric dating. All radiometric methods make use of the fact that minerals often contain minute traces of radioactive atoms, which decay at known rates into other kinds of atoms. Instruments measure the abundance of the original radioactive atoms and the abundance of the atoms that result from the decay. Since the rate of decay (half-life) is known, it is a simple matter to calculate how long the decaying must have been going on in order to have produced the measured ratio of radioactive atoms to those that have come from the radioactive decay.

When dating bony fossils, radiocarbon dating is useful only for specimens less than about 80,000 years of age. Since the radioactive form of carbon (the isotope called carbon-14, or C^{14}) decays fairly rapidly, almost none of the original atoms are left after that length of time, and it is impossible to establish a ratio. Older, less sensitive radiocarbon dating machines are reliable only with specimens under 40,000 years of age.

Much older fossils, such as those of early human ancestors and dinosaurs, can be dated by measuring the ratios of radioactive atoms that decay at much slower rates than C^{14}, and the atoms they give rise to. Thus, specimens that are many millions of years old will still contain some radioactive atoms. One of the most useful methods of such dating involves the decay of a radioactive isotope of potassium into argon, a process with a half-life of 1.3 billion years. Unlike radiocarbon dating, which dates bones directly, the potassium-argon method dates rocks, especially a type of crystal formed in volcanic eruptions. Thus, many of the dates assigned to hominid fossils are based not on the fossil bones themselves, but on layers of volcanic ash that covered the fossil before it was dug up. The potassium-argon date gives the length of time that has passed since the volcanic eruption that created the radioactive crystals in the ash. It is assumed that the bones must be older than the crystals because they were covered by the ash layer. But whereas this method is useful for very old fossils, it is unreliable for specimens less than 400,000 years old. This is because that is not enough time for the decay product, argon, to accumulate in a quantity that can be measured accurately.

There are several other radiometric dating methods, involving other kinds of atoms. All work on the same principles, but one may be preferred over the others in a given situation because of the kind of material that is to be dated. Without layers of volcanic ash, for example, the potassium-argon method is seldom useful.

See CARBON-14; HALF-LIFE; ISOTOPE.

RADIO TELESCOPE—Until the invention of this device in modern times, astronomers knew nothing about the rest of the universe except the phenomena that happened to emit visible light of an intensity strong enough to be picked up by ordinary light telescopes. As it happens, visible light is only a narrow part of a much larger spectrum of electromagnetic waves, and many objects in the sky—including objects that produce no visible light—emit electromagnetic radiation at frequencies well out of the visible range. In other words, stars, galaxies, and many other objects are sending out information in a form that our eyes cannot tune in on. Radio signals, however, can be detected with metal antennas and, through various forms of electronic processing, the signals can be focused and made into a picture something like what we might see if our eyes were capable of tuning in radio frequencies.

Among the more important discoveries of radio telescopes are vast clouds of freely drifting molecules that permeate the galaxy and which are believed to be the spawning grounds of stars. These molecular clouds, too cold to emit visible light, do emit radiation at frequencies characteristic of various molecules. The clouds consist mainly of hydrogen and helium, but with a goodly mixture of other kinds of molecules too. Radio telescopes have also detected clouds of various organic molecules (such as methane, ammonia, and formaldehyde) drifting in space, apparently produced by entirely nonliving processes. Besides these phenomena, radio telescopes have discovered some of the original energy of the Big Bang, now so dissipated that it is equivalent to a heat energy of only 3 degrees above absolute zero.

See BIG BANG; SOLAR SYSTEM, ORIGIN OF; ELECTROMAGNETICS.

RAMAPITHECUS—The oldest fossil primate that has been proposed as a hominid, or member of the human evolutionary family. Whether it actually is a hominid, however, is now in considerable doubt.

See HUMAN EVOLUTION.

RECEPTOR—Living cells are surrounded by membranes that keep their internal parts separate from the world around them. But since cells need to take in many kinds of molecules from the outside world, and have to give off other molecules, the cell membrane must contain many kinds of specialized "one-way gates" to let these molecules pass. Many smaller molecules such as water simply move through tiny holes in the membrane, in the process called osmosis. Large molecules, however, can only get through if the membrane is equipped with an appropriate receptor. Receptors are usually big molecules that the cell makes, and which either rest on the cell surface like sesame seeds on a bun, or which are partially within and partially projecting out of the membrane. The receptor waits for some other molecule to come along that happens to have the right shape to fit the complementary shape of the receptor. Some receptors are like locks waiting for the right key. If the key works, the receptor pulls the key molecule into the cell. The various kinds of cells in the body differ in the types of receptors in their membranes, thus defining the activities they are able to perform.

See MEMBRANE; OSMOSIS.

RECOMBINANT DNA—One of the most dramatic discoveries in the history of biology was the finding, about ten years ago, that the genes of any living organism can be taken apart and put together again—or recombined— at will. Genes from one species can even be spliced into the chromosomes of another, creating a new form of life with new abilities. The discovery has revolutionized biology and spawned a multimillion-dollar industry that promises (sometimes a little too hastily, as in the case of interferon which was touted as a cancer cure) great benefits for society. The new methods are loosely called genetic engineering, and the industry is called biotechnology.

Genes are sequential segments of the long-chain molecules of DNA that make up the chromosomes in every living cell. Each gene is a set of chemical instructions that tells other parts of the cell what kinds of protein molecules to assemble. Proteins are life's all-purpose molecules. Depending on their shape and chemical properties they may function as anything from the tough collagen that makes skin durable to a hormone that signals the ovaries to produce an egg, to an enzyme that digests a specific kind of food.

Recombinant DNA technology employs various methods for identifying the gene that encodes a given protein, extracting the gene from its chromosome, and splicing (or recombining) the gene into some other organism's string of genes. The gene (or genes, since more than one are often taken) may be taken from any living cell, but most often are spliced in among the genes of bacteria, and usually the common species of bacteria called *Escherichia coli,* or *E. coli,* for short. Genes, as it happens, don't care what kind of cell they're in because all genes in all living organisms work in exactly the same way. Except for one minor variation there is only one genetic code, and every cell in every creature knows it. This is one of the benefits of all life having evolved from a common ancestor. Bacteria are favored receptacles for foreign genes because they serve the chief goal of recombinant DNA technology—to produce large quantities of the protein encoded by the gene. Put a human gene into a bacterium and within days the bacterium will have multiplied into billions of identical bacteria, each bearing the inserted gene and each producing the protein specified by the gene. Huge vats of engineered bacteria may turn out large quantities of formerly scarce substances that are of use in medicine and industry. For example, interferon, a natural antivirus protein produced in the body, could not be tested properly for anticancer effects until the gene carrying the instructions for producing interferon was spliced into bacteria that then turned out large batches of this protein.

Gene splicing is not strictly a human invention. Bacteria do it all the time, shuttling various genes from one individual to another in a process of natural recombination. Bacteria also have the tools to cut DNA in specific places, insert genes into the cut locations, and splice the DNA strand back together with the new gene in it. The work is all done by enzymes, special protein molecules that seek specific points on a DNA strand and cut it in two at those points. Bacteria use these enzymes, called restriction enzymes, as a defense against invasion by viruses. When a virus penetrates into a bacterium, the invaded microbe literally chops the virus's DNA to pieces. When biologists learned this, they found there were enough different bacterial restriction enzymes, each keyed to a specific site on a DNA strand, to make it possible to pick the right combination of enzymes to "carve out" any given gene from one chromosome and splice it into any other chromosome.

Although there was once worry that gene splicing
would lead to the creation of dangerous new microbes that
could escape the laboratory, several years of experience
have satisfied nearly all of the early critics that the procedures are safe. One reason for their safety is that the
bacteria used are of a specially engineered strain that cannot live without conditions provided in the laboratory.

See GENES, HOW THEY WORK; CHROMOSOMES.

REDI, FRANCESCO, 1626–1697, Italian physician—
The first solid blow against the ancient doctrine of the
spontaneous generation of life was struck by Redi. The old
belief held that many forms of life arose *de novo* from rotting meat, swamp water, and sites of putrefaction. Maggots, for example, were thought to form in rotting meat.
Redi disproved this belief in one of the first examples of a
biological experiment with proper controls. He put meat
in several jars, sealed some with glass, and some with
gauze, and some he left open. The meat rotted in all the
jars, but only in the open jar did maggots appear. Redi
correctly surmised that they formed from eggs laid by
flies visiting the meat.

Redi was, like most biologists of his era, a physician. He
was also a poet of some small note.

See BIOGENESIS (the theory that life arises only from
other living organisms).

RED SHIFT—Astronomers use this term to talk about
how we perceive light waves that reach us from stars and
similar bodies that are moving rapidly away from us. Certain features of the light do not appear at their usual positions in the spectrum; they are shifted toward the red end
of the spectrum. These missing features that mark a red
shift are dark bands on the spectrum—places where no
light of a given frequency is reaching us because it is
being absorbed by some specific chemical element at the
light source. The pattern of these "absorption lines"
amounts to a signature for each element. If a star is moving away from Earth very rapidly, the pattern of the ab-

sorption lines in its light spectrum will be shifted toward the red end of the spectrum to a degree that has a mathematical relationship with the speed of the source. The red shift is a form of Doppler shift—the same phenomenon that shifts sound waves to higher or lower pitches depending on whether the distance between the sound source and the receiver is rapidly growing or diminishing.

See DOPPLER EFFECT; BIG BANG (red shift, which played a role in its discovery, is explained more fully there).

RELATIVITY—If you are like most readers approaching the subject of relativity, you bring to this page little confidence that you are going to understand what is written here. That's normal. However, the fact is that with just a bit of effort on your part, you will be able to grasp a lot more than you might have thought possible. Einstein's theories of relativity—arguably the most challenging and revolutionary intellectual concepts ever devised—*can* be understood in a general way by non-physicists, and without having to plow through lots of mathematics.

This is not to say that you will be able to walk up to the blackboard and chalk out some new predictions from Einstein's equations. But it is to say that you can arrive at a fundamentally new way of thinking about the nature of what we so glibly call matter, energy, space, and time; you can grasp some of relativity's seemingly paradoxical conclusions, and you can arrive at a deeper appreciation for the nature of reality—a profound appreciation that can be both disturbing and rewarding. By coming to grips with the laws of relativity you can begin to fathom the essentially mystical or religious feelings that Einstein and many others have had when contemplating the ultimate nature of reality. For relativity, beyond its purely practical implications, is also a system of philosophy, one that students of philosophy may recognize as illuminating and extending the ideas of the 18th-century English philosophers Locke, Berkeley, and Hume.

*   *   *

Berkeley, to pursue a philosophical aside, wrote: "All the choir of heaven and the furniture of Earth, in a word all those bodies which compose the mighty frame of the world, have not any substance without the mind. . . . So long as they are not actually perceived by me, or do not exist in my mind, or that of any other created spirit, they must either have no existence at all, or else subsist in the mind of some Eternal Spirit."

Einstein carried this line of thinking to its ultimate limits by showing that even space and time are forms of intuition, just as dependent on our consciousness as are concepts of color, shape, or size. Space has no objective reality except as an order or arrangement of objects we perceive in it, and time has no independent existence apart from the order of events by which we measure it.

Philosophically, this conclusion is inescapable. But it seems that if it were true, the world would dissolve into an anarchy of individual perceptions. Clearly it doesn't. We go about our lives *as if* we all perceived the same thing, and events ordinarily proceed very much in accordance with our individual perceptions. Early philosophers saw this harmony of nature as evidence of God. Scientists, often preferring to avoid supernatural explanations, invoked another kind of orthodoxy—mathematics. The universe, they said, obeys the laws of mathematics. And indeed, they seem to have been right, because they have shown time and again that they can discover new facts about the universe simply by working out equations. It is equations that led scientists inside atoms to discover quarks and out to the farthest reaches of space to search for black holes, and eventually to the Big Bang of creation itself. (Some scientists, such as Einstein, simply say that mathematical orthodoxy might just as well be called God.) But once Einstein took our understanding of the universe beyond Newton's—which holds for all practical purposes in the everyday world—the harmony of shared perceptions broke down. Only persons who could do their perceiving in the language of mathematics could see the new reality of curved space and time that runs at different speeds in different places.

*   *   *

Einstein put forth two theories of relativity—the special theory, which applies under certain restricted conditions; and the general theory, which applies to all conditions. We will deal first with special relativity.

In a nutshell, Einstein's first, or special theory has to do with objects in motion and how we perceive motion. The theory says that when we perceive an object to be moving very fast—at an appreciable fraction of the speed of light—we will also perceive certain other attributes of the object (its size and mass, for example) to be different from what we perceive them to be when the same object moves at a different speed or when it is at rest. The relativity part comes in when we realize that speed itself is relative to the observer. For example, we may watch a fast-moving object and perceive that its mass and size are altered, whereas somebody riding inside the fast-moving object may watch us and see that relative to him, we are moving very fast and our mass and size are altered. Both observations are equally true, and there is no rational way of choosing one perspective as more true or more real than the other. The same relativity effect also alters the rate at which time passes for objects moving at different speeds. Again, differing perceptions of the rate of time would be equally true, since the universe has no known standard frame of reference—no place that is truly and absolutely at rest—from which to reckon the motions.

Understanding relativity demands that we discard, or at least suspend, certain conventional ways of thinking. To do this we must start from the beginning to examine how our perceptions are formed. We will first look at some ordinary events that almost everyone has experienced, and which essentially obey Newton's laws of motion. They illustrate a pre-Einsteinian theory of relativity. Then we will examine an observation that physicists made late in the nineteenth century that seemed to violate both Newton's laws and the older theory of relativity. Finally, we will see how Einstein concluded that Newton's laws are in fact wrong, but not obviously so except for objects moving at an appreciable fraction of the speed of light. Einstein's revision of Newton's laws, published in 1905, was a new theory of relativity which he called the special theory of relativity. In correcting Newton's laws, Einstein came up with formulas that led inescapably to revolutionary new conclusions about the nature of reality—conclusions that contradict some of our most fundamental assumptions.

For example, the laws of relativity say that time is relative. It does not flow at the same rate in all places. If an astronaut flew in space at 90 percent of the speed of light and returned to Earth when his calendar told him five years had elapsed, he would find that ten years had passed on the Earth. The laws of relativity also say that the faster an object moves, the more its size shrinks in the direction of its motion, as perceived by a stationary observer. The same observer would find the mass of the moving object greater than when it was at rest. Relativity further shows that nothing can travel faster than the speed of light. (The speed of light, incidentally, is the same as the speed of any other form of electromagnetic radiation, such as x-rays, infrared waves, or radio waves.) The speed of light is an upper limit because as objects approach it, their mass approaches infinity. And most startling of all, "special relativity" asserts that matter and energy are merely two different guises of the same fundamental thing. Such is the meaning of the formula $E = mc^2$ (the amount of energy, E, contained in a given mass, m, is equal to the mass multiplied times the speed of light, c, squared) so dramatically proven in 1945 when parts of the uranium atom were converted into energy in the first atom bomb.

No matter how counterintuitive the implications of relativity, however, they command belief because every experimental test of the theory has confirmed that Einstein was right. Highly accurate atomic clocks, for example, have been flown around the world in jet planes and then

compared with twin clocks on the ground, with the result
that the moving clocks had lost the predicted amount of
time for the speed at which they were traveling. The differences were only a few billionths of a second, but the
jets were going at only about 600 miles per hour,—far below the speed of light (186,000 miles per second). The
amount of time dilation, as the effect is called, was very
close to what the relativity equations predicted (and well
within the error inherent in the measuring methods). Einstein's ideas also merit belief because it is now clear that
the strange world of relativistic phenomena is the norm in
the universe as a whole; the human-scaled realm in which
we all live and in which the effects of relativity are slight
represents only a tiny fraction of the range of phenomena
actually going on in the universe.

*   *   *

To grasp Einstein's theory of relativity it is necessary first
to look at the older kind of relativity that grew out of
Newton's laws, which involves patterns of thinking with
which almost everyone is familiar. A good example of such
thinking involves sitting in a train in a station and getting
the feeling that your train has started moving. A moment
later, however, you realize that the train on the next
track is the one that is moving, and that you are standing
still. Your perception of whether an object is moving depends on your assumption of whether another object is at
rest. If you are isolated within one frame of reference (and
cannot detect anything outside it), it is impossible to say
whether you are in motion or at rest. It was Newton who
first observed that when a person is in such a situation,
there is no way to tell whether he is moving or at rest.
You can move at 500 miles per hour in an airplane and, if
the flight is smooth, the laws of physics will operate inside
the plane as if you were parked on the runway, including
playing catch with a ball, which behaves exactly as if the
plane were on the ground, and pouring coffee, which falls
straight down into the cup. (As viewed from the ground,
however, the coffee would be seen to fall at an angle, with

a downward and forward slope in the direction of motion of the plane.) In other words, no matter how fast you are going, the laws of physics operate perfectly normally within your frame of reference as long as your motion is smooth and you are not changing speed. And these happen to be the same conditions under which special relativity applies. (General relativity extends the theory to all other conditions.)

Now let's modify the train situation slightly and say that you are in one of two spacecraft far out in space. With no Earth or other seemingly stationary object in the background, it would be impossible to say with certainty which spacecraft was moving. You could only conclude that yours was moving *relative* to the other one. An astronaut aboard the other spacecraft would reach the same conclusion with regard to his vehicle. If you could not see the other craft and had no other communication, there would be no way in which you could tell whether you were in motion or at rest.

Although it may seem that differences of interpretation of this sort are merely amusing diversions, they are actually of crucial importance to the theory of relativity. We all define "reality" as it relates to our own frame of reference, but as Einstein showed, each moving object can be its own frame of reference. There is no law of nature, no scientific principle, no mathematical formula that will point to one frame of reference as better than another in describing events that are happening.

Long before Einstein, it was obvious to many people that motion is a relative phenomenon. On Earth it is conventional to think of most motions as relative to the ground. We say that an automobile doing the speed limit is moving at 55 miles per hour. But this is only a convention, and it is easy to see that the same car's speed may be different in relation to other frames of reference. Relative to the car just behind it, the car we are speaking about may be going 0 miles per hour. Relative to one coming the other way, its speed may be 110 miles per hour.

All speeds are equally true within their frames of reference.

Newton and even Galileo understood the kind of relativity we have just been discussing, and had no trouble dealing with a phenomenon that seemed simple until 1887, when a remarkable experiment upset the Newtonian applecart. The phenomenon is easiest to consider with another airplane example. Suppose we are on the ground watching a plane go by at 500 miles per hour, and a passenger tosses a ball toward the front of the plane. Inside the plane it looks as if the ball is moving forward at, say, 30 miles per hour. Relative to the ground, however, the ball is moving at 530 miles per hour—the speed it had before being thrown plus the extra speed imparted by the thrower. If the ball is tossed from the front of the plane toward the back, it will start with the 500 miles per hour of forward speed of the plane, but then lose 30 miles per hour of speed by going in the opposite direction. The result will be a ground speed of 470 miles per hour.

In 1887 this simple understanding of relative motion collided with the results of one of history's most famous experiments—the so-called Michelson-Morley experiment to measure the Earth's motion. This is the conflict that Einstein would resolve with his new theory, but again, to understand how he did this, a little background is necessary.

The Michelson-Morley experiment was based on the assumption that light was a wave and that, like all waves, it consisted of a disturbance in a medium that propagated it, just as sound waves are propagated in air. No propagating medium was known for light waves, but virtually all physicists agreed that one had to exist. It was called ether (from the same root that gives us "ethereal," and not to be confused with the anesthetic). It was believed that light waves were vibrations that moved through the ether at a fixed speed, just as sound vibrations do through the air.

Albert Michelson and Edward Morley, both Americans, got the idea that the ether could be considered a stationary frame of reference that permeated the universe and that it offered a means of detecting the "true" motion of

the Earth. They assumed that light traveled at a constant
speed through the ether (as does any wave in its medium),
and believed that if the light were moving through the
ether in the same direction as the ground were moving
under it, the ground speed of the light would be clocked as
slightly slower than if the Earth were standing still. (In
other words, as the light moved at its fixed pace through
the ether, the Earth would be moving in the same direction and, relative to the Earth, the light would seem to go
slower than normal.) Michelson and Morley assumed that
the true speed of light (its airspeed, in a sense) could be
measured in a beam of light aimed perpendicular to the
Earth's direction of motion. They devised an experiment
in which a single light beam was split, sent in two directions (parallel to the ground's motion and perpendicular to
it), and then rejoined in a way that allowed the detection
of extremely small differences in the clocked speeds of
each half of the split beam. The method was reasonable
and the equipment was precise enough to detect the difference they were looking for.

But Michelson and Morley, like the entire physics community, were unprepared for the result: There was no difference in the speeds of the two halves of the split beam.
Physicists were faced with two interpretations of this: either there was no ether (which meant that light was a
wave traveling in nothing!), or Copernicus and Galileo
were wrong (which meant that the Earth really did sit
immobile at the center of the universe!). Since there was
abundant other evidence that the Earth moves, physicists
had to conclude that ether did not exist.

The conclusion was troubling because it conflicted with
one of the fundamental principles of the old relativity, the
idea that the speed of an object can differ, depending on
the reference frame of the observer. Imagine a situation
with two cars moving together down the highway, one
going 55 miles per hour and the other going 54 miles per
hour. An observer in the slower car would clock the faster
one as gaining ground at the rate of 1 mile per hour. The
situation with light, however, is very different. Imagine a

beam of light hurtling along at 186,000 miles per second and someone in a spacecraft flying parallel and in the same direction at 185,000 miles per second. Ordinary Galilean relativity would say that the astronaut should clock the light as gaining speed at the rate of 1,000 miles per second. Einsteinian relativity, by contrast, says that the astronaut will clock the light at a full 186,000 miles per second.

In other words, Einstein saw a law of nature in the results of the Michelson-Morley experiment: The speed of light is the same for all observers, regardless of their own motion; lightspeed is a universal absolute. It is, of course, most peculiar that light should be like this. It is totally contrary to expectations based on observations of every other thing that can move. The constancy of lightspeed is a deep enigma of nature, but it is true, and Einstein's great insight was to take it as a fact and run with it.

Accordingly, the special theory of relativity basically amounts to a revamping of Newtonian physics such that when you work out the formulas, the relative speed of light always stays the same. It never changes relative to anything else, although some other things do. They must change if lightspeed is to stay constant in the equations. Mass, space, and time all vary depending on how fast you move. The faster you go, the greater your mass, the smaller the space you occupy, and the more slowly time passes. These effects are most pronounced when objects are moving at an appreciable fraction of the speed of light. As previously mentioned, an astronaut traveling at 167,000 miles per second (about 90 percent of lightspeed) for five years (according to his watch and calendar) would return to Earth and find that ten years had passed there. At 99.99 percent of lightspeed, the astronaut could fly for only six months by his reckoning and return to find that fifty years had passed on Earth.

While the most dramatic effects of the law of relativity occur at such high speeds. relativistic changes still happen in the day-to-day world, although we never notice them unless we are using very sophisticated equipment. People who fly once around the world toward the east (it makes a difference whether you go with or against the Earth's rotation) come home about 40 billionths of a second (40 nanoseconds) younger than if they had stayed at home.

Though relativity's warpings of time, space, and mass still strike most people as strange or even impossible, physicists have gotten used to them—after all, it has been about eighty years since they were introduced—and some encounter the phenomena of relativity quite regularly in their experiences. This happens most often inside particle accelerators, where subatomic particles such as protons are accelerated to velocities nearly that of light. When such a proton hits a target—perhaps a sheet of metal—it blasts a hole in it as if it had an enormously increased mass. Some of the energy of the impact is spontaneously converted into particles of matter—usually unstable ones that fly out of the impact zone and disintegrate (turning back to energy) almost instantly. Physicists have found out what types of particles are formed in such a collision and how long they live before disintegrating. They have also noticed a startling confirmation of relativity theory: that the faster the particle moves as it flies out of the impact zone, the longer it lives before disintegrating. Time, for that particle, is stretched out.

The relativistic effects discussed so far are those that emerged from Einstein's special theory, published in 1905. It was special because its calculations applied only to objects moving in a straight line without changing speed. But Einstein knew that such a limitation meant the theory could not explain what happens when objects are accelerating, slowing down, or moving on a curved path. And these three circumstances include most forms of motion in the tangible world. To extend the insights of relativity to these cases, Einstein knew he needed a general theory. His search took the next eleven years, but in the end it yielded a theory that many physicists consider the most awesome and elegant intellectual achievement of all time. A proper review of the theory, unfortunately, is beyond the scope of this book. Suffice it to say that the general idea of the relativistic phenomena revealed by the special theory is preserved, with the addition of a new way of looking at gravity—the force that causes acceleration and deceleration, and which curves the path of moving objects such as planets.

To fit gravity into his theory, Einstein found a way to say that gravity is not a force at all but an artifact of the way in which we observe objects moving through the four dimensions of space-time (the usual three dimensions of space plus the one dimension of time). He was able to take this approach when he realized that there was no way to tell the difference between the effects of what we call gravity and the effects of acceleration.

To explain the equivalence of gravity and acceleration, Einstein envisioned an elevator whose cables had broken, releasing the elevator and its occupants to fall freely. In such a situation, the people inside would no longer feel any pull from gravity. Their situation would be like that of astronauts in orbit, who are "weightless" because, like the occupants of the elevator, they are in free fall around the Earth. Assuming that the occupants could not look out of the elevator, there would be no way for them to tell the difference between plummeting to the ground and either being in orbit or flying among the stars in outer space. Then Einstein envisioned what it would be like if somebody grabbed the cables and pulled the elevator upward. Einstein noted that aside from averting disaster, the pull—if it accelerated the elevator upward at just the right rate—would make it feel as if gravity had come back into play inside the elevator. Even if the elevator were deep in outer space, there would be no way to tell the difference between acceleration and gravity.

This equivalence allowed Einstein to write equations describing motion through space as if gravity were not a force, but instead, as if massive bodies (such as planets) caused a curvature in space-time—a slope along which objects could slide into the massive body. The situation is sometimes represented in a diagram showing a planet or the sun as a ball at the bottom of a huge, curved funnel.

The funnel represents a two-dimensional version of the surface of space-time. Anything moving toward the ball in what seems to start as a straight line actually rolls along the funnel's curved surface and, in effect, falls into the hole. Curved space-time warps the object's path just as a curvature in the green of a golf course warps the path of a ball that is otherwise moving straight. In reality, every object with mass is surrounded by a nearly infinite number of funnels converging on the object from all directions.

Odd as it seems, the curved spacetime of general relativity has been confirmed in experiments. Einstein's equations, for example, successfully explained an oddity in the orbit of Mercury that Newtonian physics could not explain. (Mercury, being so close to the sun, maintains an orbit along the steeper slopes of the funnel and is more noticeably affected by the sun's gravity than are more distant planets.) General relativity also correctly predicted the amount by which starlight is bent as it passes near the sun. This was measured when, during a total eclipse, astronomers looked for light grazing the sun's disk as it traveled from a star of known position. The star's position appeared shifted to one side—precisely the effect that would occur with bending of the path of light emanating from the star.

Although relativity theory may not seem greatly consequential in the world around us (it isn't), it has passed every experimental test devised for it, and it has led to some of the most fascinating predictions about the universe at large. It has, for example, led to the prediction that black holes exist. These are regions of spacetime so sharply curved that any light emerging from them immediately curves back into its source. Relativity theory has also led to mathematical descriptions of what must have happened in the earliest moments of the Big Bang in which the universe was created.

See EINSTEIN; GRAVITY; SPACE-TIME; BLACK HOLE; BIG BANG.

REPLICATION—This term has two main uses in science. The broader use refers to the scientific tradition that any experiment must be replicated, or repeated by another researcher, before it can merit belief. In this context, replication is a method of minimizing the chance of error or misinterpretation. The narrower meaning applies to the process by which the DNA molecule of the chromosome makes a duplicate of itself.

See Part 1, What is science?; Double helix.

RETROVIRUS—This is a special type of virus that has both good and bad faces. Some retroviruses cause cancer, but modified versions of these viruses (without the cancer-causing genes) are being used to develop cures for genetic diseases. As the name suggests, a retrovirus is a kind of backwards virus. Like all viruses, retroviruses consist of genes wrapped in a protein coat. And like other viruses, the retrovirus cannot, by itself, reproduce; it must invade a cell and commandeer the cell's metabolic machinery. The genes of ordinary viruses are made of DNA and, once in a cell, directly begin to command the cell to make new viruses. The host cell is soon literally bursting with new viruses, which break out, killing the cell, and go on to infect new cells. Retrovirus genes, on the other hand, are made of RNA, a molecule closely related to DNA. Instead of setting their genes directly to work on the host cell, retroviruses produce an enzyme that makes a DNA copy of their RNA (the reverse of the usual sequence), and then splices this copy into the chromosomes of the host cell. Once this is done, the virus's genes remain a part of the cell's genetic heritage, and are passed on to the daughter cells that result from cell division. These new host cells make new retroviruses, which emerge and infect new cells, but are not killed in the process.

See Virus; Gene therapy; Genes, how they work.

RIBONUCLEIC ACID (RNA)—A long-chain molecule similar to DNA but with a different purpose. The job of DNA is to preserve a cell's genetic instructions safely within the nucleus, and to duplicate itself so that it can pass on a complete set of these instructions to each of the two cells that result from cell division. There are two main kinds of RNA, messenger RNA, or mRNA, and transfer RNA, or tRNA, each with a specific job. The job of messenger RNA is to transcribe the genetic message from DNA and carry it out of the nucleus to the cellular machinery that can read it and follow its instructions. The machinery that follows the instructions is the ribosome, which makes the protein molecules specified by DNA. To do so, it must string together the subunits called amino acids. This is where transfer RNA comes in. Transfer RNA is a short molecule that grabs a specific amino acid and transfers it to the ribosome. Depending on a specific code in the messenger RNA, a particular kind of transfer RNA (one of which exists for each of the twenty kinds of amino acid) will come to the ribosome, drop off its amino acid (which binds to the chain that will become a protein), and drift away to find another molecule of the same amino acid. Following this, another transfer RNA, carrying its particular amino acid, will be called by the next specific code in the messenger RNA, and will come and add its amino acid to the lengthening protein chain.

See DNA; RIBOSOME; GENES, HOW THEY WORK.

RIBOSOME—A "machine" inside a cell that reads the genetic code (brought by messenger RNA from the DNA in the nucleus) and assembles the protein molecule that is called for by the code. Messenger RNA carries its message as a linear sequence specifying, one at a time, the kinds of amino acids that must be linked together to make a given kind of protein. There are twenty kinds of amino acids, and their sequence determines the shape and function of the final protein molecule. The ribosome moves along the messenger RNA molecule, reading the sequence of instructions it contains. If the next amino acid needed is arginine, for example, the ribosome waits until an arginine molecule happens along (there is a veritable soup of mixed amino acids inside the cell) and positions it to bind chemically to the previous amino acid. Proteins typically have a sequence of hundreds to thousands of amino acids. As the amino-acid chain emerges from the ribosome, it begins curling and folding into the characteristic three-dimensional shape of the protein, which is determined by the attractions and repulsions between atoms at various points along the protein chain.

See PROTEIN; GENES, HOW THEY WORK; DNA.

RNA–See RIBONUCLEIC ACID.

RUTHERFORD, ERNEST (LORD), 1871–1937, New Zealand physicist—Rutherford was one of the founders of the modern science of atomic physics. His chief contribution was to develop—some three-quarters of a century ago—a model of the internal structure of atoms that is very close to the modern view. He deduced that atoms consist of a small, dense nucleus having a positive electrical charge, surrounded by a swarm of negatively charged electrons.

Rutherford was born in the small New Zealand town of Spring Grove, near Nelson. He was the fourth of twelve children born to a wheelwright and his wife who had emigrated from England. Young Rutherford proved a bright boy, and his parents endured hardship to send him to a private school. In the course of a college career in physics and mathematics, Rutherford performed experiments on the then newly discovered radio waves that won him a scholarship to Cambridge University.

He arrived in England in 1895 and began to work at the university's famed Cavendish Laboratory under J. J. Thomson, a pioneering physicist who had discovered the subatomic particles called electrons. Later that year he heard of the discovery by Wilhelm Röntgen in Germany of a mysterious ray that could penetrate glass and even human flesh to make a kind of shadow picture on photographic film. (The rays were emitted by specially built glass tubes through which electrical currents were passed.) With suitable mystery, the rays were dubbed x-rays. Rutherford experimented with x-rays and found that they ionized the air through which they passed—or, in other words, that they split molecules of the gases in air into electrically charged atoms. Rutherford found ways to measure the speed with which the rays traveled and how fast the ionized atoms recombined into electrically neutral molecules. These methods were to prove invaluable in the following year, when Henri Becquerel in France discovered that a mineral called uranium emitted some kind of mysterious rays that, like x-rays, could fog photographic film.

Rutherford obtained some uranium and applied his measuring methods to the rays it emitted. He learned that uranium rays came in two forms, both of which were different from x-rays. He named them simply alpha and beta rays, and devoted the next few years to trying to learn what they were. He showed that alpha rays had very little penetrating power but were strong ionizing agents. Beta rays, by contrast, were highly penetrating but had little ionizing ability. In 1903 Rutherford, having since moved to McGill University in Montreal, showed that alpha rays could be deflected by electrical and magnetic fields, with the direction of the deflection revealing that they carried a positive electrical charge. Rutherford's work on alpha rays took most of a decade, but along the way it became clear to him and his collaborator, Frederick Soddy, that the rays were actually particles emitted by atoms that were, for some unknown reason, unstable. Alpha particles, he found, were the same as helium atoms without their electrons, and beta particles were simply individual electrons. Certain kinds of atoms, Rutherford realized, were spontaneously casting out some of their components and transforming themselves into different kinds of atoms. Some atoms, it seemed, contained helium atoms or were made of helium atoms stuck together. The theory met with considerable resistance because it smacked of the magical notions of alchemy, with its efforts to transmute "base" metals into gold. Eventually, however, the idea caught on, and stands today more or less as Rutherford stated it.

In 1907 Rutherford moved back to England, this time to the University of Manchester where, along with Hans Geiger, he invented the radiation-measuring device known today as a Geiger counter.

Rutherford's greatest contribution to science was in formulating what is known as the nuclear model of the atom. Through most of the nineteenth century, atoms, if they were believed in at all, were thought of as solid, indivisible balls of matter. Then, late in the century, Thomson's discovery of the electron had led to the view that all atoms contained electrons. Thomson suggested that atoms consisted of a ball of matter permeated by electrons, rather like raisins in a cake. Since electrons had a negative charge and atoms were usually neutral, Thomson supposed the dough of the cake to have a positive charge that was neutralized by the negative raisins (electrons).

Some of Rutherford's experiments in shooting alpha particles through other substances led him to a bold revision of Thomson's atomic model. The insight emerged from an experiment in which he was beaming alpha particles at sheets of photographic film protected against light. Normally the beam produced a sharp-edged image on the film, a shape corresponding to the shape of the device emitting the particles. But when Rutherford placed sheets of metal foil between the emitter and the film, the image turned fuzzy. Some of the particles were being deflected by the foil, their path bent. Further experiments revealed that the foil was not only deflecting some particles, but that it was actually bouncing a few back almost toward the emitter.

This, Rutherford said, was "quite the most incredible event that has ever happened to me in my life. It was almost as incredible as if you fired a 15-inch shell at a piece of tissue paper and it came back and hit you."

Clearly, Rutherford reasoned, the metal foil could not have been a perfectly even barrier to the passage of alpha particles. The metal atoms must have been formed in such a way that the vast majority of particles could pass straight through, that a few were bent in their paths, and that a few were actually bounced back without penetrating at all. The explanation Rutherford thought of was that atoms consisted mostly of empty space (which would allow most alpha particles to pass through the foil unhindered), but that they had all of their positive charge concentrated in one spot. He calculated that the only way in which an alpha particle could be bounced back toward the emitter was by encountering a powerful enough concentration of positive charge to bounce it back. (Since alpha particles are positively charged, they would be repelled by encountering another positive particle.) The only way an atom could have enough positive charge to do this trick would be to have all of the charge concentrated in one place—the place through which the rare alpha particle might try to pass. Rutherford thought of the positive charge as a kind of atomic core, around which were orbiting the electrons. He soon called the core a nucleus.

There was only one problem with Rutherford's model. The laws of classical physics say the electrons should quickly lose energy and crash into their nuclei. It was Niels Bohr who, just a few years later, explained why this does not happen. Still, the Rutherford model of the atom, as explained by Bohr and as modified somewhat in later years, still stands.

See ATOM THEORY; BOHR.

SALAM, ABDUS, 1926– , Pakistani physicist. See GRAND UNIFIED THEORIES.

SCHLEIDEN, MATTHIAS, 1804–1881, German botanist. See CELL THEORY.

SCHWANN, THEODOR, 1810–1882, German biologist. See CELL THEORY.

SCIENTIFIC METHOD—A set of procedures and principles that scientists follow when doing research, so as to minimize their chances of being fooled by the results of an experiment. There is no one set of procedures. This topic is discussed at length in the Introduction to this book, entitled "What Is Science?"

Also see the biographical sketches of scientists who helped develop the underlying ideas: ARISTOTLE; BACON, FRANCIS; BACON, ROGER; GALILEO; DESCARTES; LAVOISIER; HELMHOLTZ.

SEAFLOOR SPREADING—One of several phenomena, including continental drift, that are part of the broad theory of plate tectonics. The New World and the Old World are moving apart at the rate of one or two centimeters a year because the Atlantic floor is spreading from a mid-ocean ridge, where magma wells up to fill the gap left by the spread.

See PLATE TECTONICS; WEGENER (who proposed the theory).

SILENT DNA—Regions of chromosomes that seem to contain no recognizable genes. The molecule of which these regions are made is DNA, but it comprises a sequence of bases (or nucleotides) that do not hold the code for any protein or serve any known regulatory role. Some silent DNA consists of the same nonsense sequences repeated many times over. By some accounts, there is more silent DNA than expressed DNA.

See GENES, HOW THEY WORK.

SIMPSON, GEORGE GAYLORD, 1902–1984, American paleontologist. See MODERN SYNTHESIS.

SOCIOBIOLOGY—A relatively new branch of evolutionary biology that seeks to explain how evolution could have produced behavior patterns that seem to be instinctive. The vast majority of sociobiology research is done on animals—the division of labor in insect colonies, the nesting habits of birds, the communications systems of monkeys, and hundreds of other examples. Sociobiologists assume that such rigid behavioral patterns are something like reflexes, and that they constitute behaviors permanently programmed into the structure of animal brains. These brain structures and their resulting behaviors, they say, evolved through the same mechanisms that shaped the more obvious physical features of animals.

Many sociobiologists suggest that some human behavior patterns may have evolved in the same ways. Such ideas have elicited strong opposition from those who believe human behavior is immune from evolutionary forces because of the nature of the human mind and the possession of free will. Opposition to sociobiology is sometimes strongly ideological, but among opponents of a more objective scientific bent, there is a continuum of opinion about which behaviors are innate and which are learned.

One of the central problems of sociobiology has been explaining the evolution of so-called altruistic behaviors that seem to diminish, not increase, an individual's reproductive success.

See ALTRUISM; EVOLUTION, THEORY OF; FITNESS; NATURAL SELECTION.

SOLAR SYSTEM, ORIGIN OF—Astronomers generally agree that the components of our solar system—the sun, planets, moons, asteroids, and others—were formed at about the same time from one vast cloud of gas and dust that was swirling about a common center. This was about 4.6 billion years ago, when the universe was already perhaps 8 to 15 billion years old. Galaxies—themselves formed from larger clouds—are full of smaller clouds that seem to drift with little change until hit by a shock wave of the sort that exploding stars, or supernovas, produce. This shock wave can squeeze together small concentrations of gas that, by becoming a region of greater mass, exerts a stronger gravitational pull on the gas around it. As more gas is pulled into the center, the stronger center pulls in still more gas.

Eventually, say astronomers, most of the cloud from which the solar system arose drew toward the center, where the growing clump of matter—mostly hydrogen and helium—became steadily stronger in gravity. When the mass became great enough, the internal pressure ignited the nuclear fires that turned the mass into a huge, long-lived hydrogen bomb or, in other words, a star. As the sun was forming, its gravity and related physical forces pulled the cloud ever more tightly into a revolving disk shape with the sun at the center. Farther out in the disk, clumps of matter were forming. The largest of these became planets, and around the larger planets, smaller clumps of matter became moons.

Mixed in with the hydrogen and helium were smaller amounts of other, heavier atoms—including iron, nickel, and silicon. Most of these heavier atoms had been formed inside other stars that had long since exploded, casting the molecules adrift. As the sun grew at the center of the swirling disk, its heat and gravitational pull began to change the composition of the disk. Denser atoms and molecules (metals and silicates, for example) sank closer to the center, while lighter molecules (hydrogen, helium, neon, and others) "floated" higher in the disk, or farther out toward its periphery.

As the solar system formed, different atoms were able to condense into solid particles at different distances from the heat of the center. The atoms that make up stone and metal, for example, condensed relatively close to the nascent sun, forming Earth and the similar so-called terrestrial planets—Mercury, Venus, and Mars. Farther out, where it was cooler, the lighter molecules condensed into the "Jovian planets," Jupiter, Saturn, Uranus, and Neptune, the so-called gas giants. These planets have relatively small rocky cores and consist mainly of liquid hydrogen. Some have ices of water, ammonia, and methane near their core. (If Jupiter were larger, its interior pressure would be enough to start a thermonuclear reaction in the hydrogen that makes up so much of this planet, turning Jupiter into a star. Then we would be living in a binary star system of a sort that seems to be fairly common around our galaxy.) Pluto, a relatively tiny planet

THE SOLAR SYSTEM IN SCALE

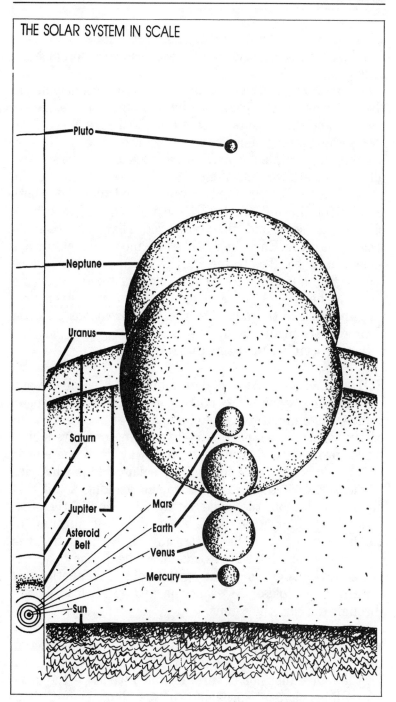

with an oddly shaped orbit, does not appear to have formed as a planet, and may be an escaped moon of Neptune.

One powerful source of evidence for this scenario lies in the fact that all the planets orbit the sun in the same direction, and their orbits (except Pluto's) are very nearly in the same plane. Moreover, all of the planets' orbits, though elliptical, are nearly circular (except Pluto's). And all the planets (except Venus) also rotate or spin, in the same direction. Venus may have started like the others but, under the influence of Earth's gravity, it may have slowed its spin and then reversed slightly, becoming locked onto the Earth much like the moon, which always shows the same face to the Earth. Venus shows the same face to Earth every time it comes close. Also evidence of a common heritage of the solar system from an original swirling disk is the fact that the various moons all orbit their planets in the same direction.

See EARTH, ORIGIN OF; BIG BANG.

SOLAR WIND—As the thermonuclear fires of the sun continue to explode like billions of successive hydrogen bombs, huge quantities of atomic particles are blasted out of the sun. These particles—hydrogen ions, free electrons, protons, and neutrinos—fly away from the sun in all directions at great speeds. This solar wind is what rips particles out of the head of a comet, making a tail that sweeps back from the comet in the direction of the solar wind. The solar wind is also the cause of auroras, or northern lights. As the electrically charged particles of the wind near the Earth's magnetic field, they are deflected and made to move along the lines of magnetic force toward the poles. As the charged particles shower through the air, they glow.

SPACE-TIME—Though it sounds like a science-fiction concoction, space-time is quite real. It is Einstein's term for the four-dimensional world in which we live—the usual three dimensions of space and the one dimension of time.

Space-time is a necessary concept if we are to think rationally about events, or transitory occurrences. To identify any point, we must use the three dimensions of space. But to identify an event we must use not only the dimensions of space to locate the event, but a dimension that tells us *when* the event happened. It sounds trivial but it is fundamental when dealing with the truths of Einstein's relativity theory, which reveal that none of the four dimensions is absolute. Depending on how fast an object moves, the dimensions of space can stretch or shrink, and the passage of time can speed up or slow down. These effects are appreciable only when objects travel at a goodly fraction of the speed of light.

See RELATIVITY.

SPALLANZANI, LAZZARO, 1729–1799, Italian physiologist—Aspects of Spallanzani's career sound surprisingly modern for a scientist working 200 years ago. He transplanted organs between various lower animals. He developed methods for the artificial insemination of dogs. He experimented with the ability of animals to regenerate lost limbs, and found that the "higher" an animal is in the evolutionary sense, the less regenerative ability it has. He found that animals produce different types of digestive juices depending on what they have just swallowed.

So it may come as a letdown to learn that when he studied the reproductive role of the male's semen, Spallanzani thought the sperm were merely parasites infecting the male. He decided semen was necessary to start a new individual growing, but thought that its active ingredient was in its liquid portion. The role of sperm, incidentally, was not understood until around 1870.

Spallanzani is probably best remembered, however, for his work on disproving the theory of spontaneous generation. This was the old idea that some forms of life can arise by the spontaneous transformation of nonliving matter. The old theory had been disproven for larger organisms by Francesco Redi, another Italian, a century earlier, but people still believed that microorganisms formed in stagnant pools of liquid. Spallanzani's controlled experiments showed that unless air can reach a nutrient broth, no microbes grow in the broth. Today we understand that such growth does occur because the dried spores of microbes float in the air as dust, and revive only when they fall into a suitably moist environment.

Spallanzani was born to a well-to-do family, his father having been a distinguished lawyer in Modena. He received a classical education at a Jesuit college and was invited to join the order, but declined and went elsewhere to study law. Eventually he was ordained, but had become too interested in science to practice either law or the priesthood. He became a professor of physics at the University of Modena, and devoted most of his time to research. In addition to the studies already cited, Spallanzani also did research on the mechanics of skipping stones across water, various forms of microscopic life, the circulation of blood through the lungs, the electric charge of the torpedo fish, and the sense organs of bats.

See BIOGENESIS (the theory that life arises only from other living organisms).

SPECIES—Despite its being such a widely used term, "species" has no ironclad definition in biology. The usual textbook definition is that a species—the smallest commonly used taxonomic classification—constitutes a group of physically similar individuals capable of interbreeding and producing fertile offspring. (The next largest classification is the genus, which may include several species.) Thus a horse and a donkey represent separate species because their offspring is an infertile mule, while a palomino and a quarterhorse are the same species because their offspring is fertile. Usually this definition of species is useful, but there are plenty of cases in which it breaks down. For example, all domestic dogs belong to the same species, *Canis familiaris*. Yet if a Chihuahua and a Great Dane were newly discovered, biologists would unhesitatingly place them in different species and perhaps even in different genera (the plural of genus) simply because they look so different. But the fact is that they can interbreed, as can all domestic dogs. So far, so good. But as it happens, dogs can also interbreed with wolves, *Canis lupus*, and with coyotes, *Canis latrans*. Biologists probably should have put dogs, wolves, and coyotes all in one species. The definitions break down even further among plants, where it is fairly common to cross one genus with another (for example, wheat with rye) to produce a wholly new genus.

The differences that lead scientists to assign organisms to different species are not always easy to see. Sometimes it takes an expert to spot them—although in most cases the organisms have no difficulty themselves. Sometimes the differences are impossible to see with the naked eye. There are, for example, frogs in some South American ponds that look exactly alike, but which when examined genetically are found to have more differences than similarities. These identical-looking species are as far apart in evolutionary terms as people and cows, but all of their differences are confined to body chemistry. Since the differences are represented in the genes, the different frogs are as unable to interbreed as are people and cows. They look the same, apparently, because they live in exactly the same habitat and are subject to the same environmental forces.

See GENUS; EVOLUTION; TAXONOMY; LINNAEUS (the scientist who devised today's system of taxonomy).

SPERM—Short for spermatozoon, the sex cell contributed by the male, which fertilizes the female's ovum. In most species, but not all, it has a tail-like flagellum that whips and undulates to help the sperm swim toward the ovum. Though the ovum is relatively passive, it contains nearly all the cellular machinery that will do the work of the newly conceived organism. The sperm is little more than a swimming bundle of paternal genes.

See OVUM.

SPONTANEOUS GENERATION—The idea, once widely held, that life arises spontaneously in various

places, such as the garbage heaps that bring forth maggots and the junkyards that seem to spawn rats. It
took many years of diligent experimentation to prove the
contrary law of biogenesis, that life comes only from
life.

See BIOGENESIS; SPALLANZANI; REDI (the two scientists who established the law of biogenesis.

STAR—A star is a battleground between two opposing
forces—the gravity exerted by its great mass, which
would cause its matter to collapse toward the center, and
the outward pressure of a sustained thermonuclear reaction that is fusing hydrogen atoms into helium atoms and
releasing energy in the process just like a hydrogen bomb.
Most stars go through similar life cycles with variations
that depend mainly on the amount of matter (almost all of
it being hydrogen) that originally went into them. As long
as its fuel holds out, the star shines. When it is gone, the
smallest stars briefly swell to become many thousands of
times larger than they were (becoming red giants) and
then cool gently down to a cinder called a white dwarf.
When larger stars exhaust their nuclear fuel, they explode
briefly into a supernova and then collapse into a dense,
spinning core that may still emit a small amount of energy
as a pulsar. If the amount of matter remaining is very
large, the core may continue collapsing under its own
gravity and become a black hole.

Stars come in many sizes. The sun is a typical middle-
aged star that, in about five billion years, will swell into a
red giant (its size may engulf the orbits of Mercury,
Venus, and possibly the Earth) and then collapse into a
white dwarf no bigger than Earth. The universe is already
populated with many stars in the different stages of their
lives, from dwarfs to giants. If the sun were the size of a
grain of sand, the red giants would range from the size of
a plum to that of an orange. On the same scale, the super-
giant stars would be bigger than beachballs. Stars also
vary tremendously in surface temperature and luminosity.
The hottest stars, blue-white in color, have temperatures

up to 100,000 degrees Celsius, while the coolest, which are red in color, are only around 2,000 degrees. The hottest stars, usually called white dwarfs, however, are so small that the amount of energy coming from them is much less than that coming from the cool red giants millions of times larger. Likewise, some stars are thousands of times brighter than others. Canopus, one of the brightest, and a common star for spacecraft to fix upon for navigation, shines some 80,000 times more brightly than the sun. One of our nearest neighbors, Wolf 359, is only one thirty-thousandth as bright as the sun.

We live in one of the sparsest parts of the galaxy, far from the dense core. As a result, only eight stars are within 10 light-years of Earth. The closest is Proxima Centauri, at 4.3 light-years away.

Stars come in many ages and lifespans as well. Some stars were formed shortly after the start of the Big Bang; others are being formed right now. If the universe is about 20 billion years old (by some estimates it is only 13 billion years old), our galaxy is about 10 billion years old and our sun only 5 billion years of age. (Our sun, like most stars, formed as part of the same process that created an entire solar system.) The bigger the star, the shorter its lifespan. Although stars generally follow the same life cycle, the speed with which they go through it depends on the amount of matter they contain. While the biggest stars may die in a relatively brief 10 million years, the smallest ones can probably keep shining for hundreds of billions of years. Our sun appears to be only about halfway through a 10-billion-year lifespan.

See SOLAR SYSTEM, ORIGIN OF; BLACK HOLE; PULSAR; SUPERNOVA.

STATISTICAL SIGNIFICANCE—When someone says the results of an experiment were statistically significant, he is saying that the effect observed is highly unlikely to have been the result of chance alone. One determines statistical significance by performing one or more mathematical analyses of the numbers involved in an experiment—the sizes, or number of individuals included in the samples in the experimental and control groups, the magnitude of the difference in how the groups fared, and such.

Tests of statistical significance are especially important in medical research, where, for example, it is common that patients in both an experimental and a control group will show improvement even though only the experimental group receives a real drug. (Expectations and other psychological factors often lead to reports of improved condition among those whose treatment, unbeknownst to them, is a sham or a placebo.) If, for example, 60 percent of the patients in the experimental group get better, as compared with only 40 percent of those in the control group, does this mean that the drug works? Not necessarily. Chance alone might produce such a skewed result. After all, if you toss a coin ten times, you don't always get five heads and five tails. If few patients were tested, chance plays a big role in the proportion of outcomes, either beneficial or nil. The more patients in each group, the lower the likelihood that chance alone can produce such a difference.

A mathematical test of statistical significance can tell the scientist just what the odds are that his results were produced by chance alone. This is usually expressed as the odds that chance produced the measured result of the experiment. A general rule of thumb is that experimental results are considered safely out of the realm of chance (statistically significant) if the odds are less than 5 percent that chance alone could have produced the measured result. This is expressed as "significant at the 0.05 level of confidence." In some cases, researchers will insist on even greater levels of significance before accepting the measured result as real—they may, for example, want the results to be significant at the 0.01 level of confidence.

See PART 1: WHAT IS SCIENCE?

STRONG FORCE—One of the four fundamental forces in nature, the strong force holds together the particles of the atom's nucleus—neutrons and protons. The strong force—or strong interaction, as it is also called— is about one-hundred times as strong as electromagnetism, one of the other three fundamental forces. (The other two are the weak force, which is involved in atomic decay, and gravity.) The strong force must be strong in order to bind protons, which have like charges and naturally repel one another. The strong force is transmitted, or communicated by the exchange among neutrons and protons of particles called virtual mesons.

See ATOMIC STRUCTURE; BIG BANG; GRAND UNIFIED THEORIES.

SUN—A star that condensed out of the same dust cloud that gave birth to the Earth about 4.5 billion years ago.

See STAR; SOLAR SYSTEM, ORIGIN OF.

SUNSPOT—A dark patch on the sun's surface. Sunspots look dark only because they are cooler—4000 degrees Celsius as compared with the 6000 degrees of the rest of the sun's surface. If a sunspot could be removed and put off into space, it would shine as brightly as a full moon. Sunspots measure anywhere from a hundred miles or so across to several times the diameter of the Earth. They are believed to result when regions of powerful solar magnetism, formed deep inside the sun, reach the sun's surface and, in effect, trap matter, holding it near the surface longer and allowing it to cool more than usual. The origin of the magnetic anomalies is poorly understood, but a leading hypothesis suggests that they result from warpings of the sun's magnetic field. Since the sun is fluid, different parts of it rotate (on the sun's axis) at different rates, bending the lines of magnetic force into spirals through the sun's interior. Sometimes, so the idea goes, the spiraling lines warp close to the sun's surface, causing spots.

There are almost always a few sunspots in evidence, but the number rises and falls in an eleven-year cycle. This is the result of the periodic reversal of the sun's magnetic poles, which happens every eleven years. Since it takes two reversals to bring the poles back to their original position, astronomers sometimes speak of a twenty-two-year solar cycle. There is growing evidence that cycles of sunspot activity have an influence on the Earth's weather. It seems that peaks in the numbers of sunspots correspond to periods of warmer summers and winters on Earth. This link is supported by the observation that between 1645 and 1715 astronomers saw very few sunspots at all, while these same decades saw such a sustained cold climate in Europe that the period is now known as the Little Ice Age.

SUPERCONDUCTIVITY—A phenomenon occurring within certain metals at temperatures close to absolute zero, and in which all resistance to electrical conduction suddenly disappears. A current induced in a ring of wire at this temperature would continue to flow in a circle forever.

See ELECTRICITY.

SUPERNOVA—It is the fate of every star to die once it has exhausted its nuclear fuel (hydrogen and helium). Stars with little mass die gradually, but those with more mass experience a cataclysmic demise. As their fuel is exhausted and their nuclear fires die down, their remaining outer matter begins to collapse. The increasing pressure of this collapse creates heat, which causes a sudden flare-up in the rate of nuclear burning, and the star explodes into renewed nuclear fires. For a few seconds it may shine as brightly as an entire galaxy. Then it may give as much light as 200 million suns for several weeks. It is this incredibly bright light that we call a supernova.

About every fifteen to fifty years, astronomers calculate, one star in our galaxy should "go supernova," creating a point of light so bright it would easily be seen in the daytime. However, the last such supernova was in 1604, which either means that we are long overdue for one or that the calculations for the occurrence of supernovas are wrong. The only supernovas seen since the invention of the telescope in 1609 have been in other galaxies.

Once the great flash of light of a supernova dies down, having burned up all the remaining nuclear fuel in the star, the remaining mass of the star (all of the heavier elements formed by fusion) collapses into a solid mass. Within seconds, the atoms pile up at the center, reaching densities in the billions of tons per cubic inch. Electrons are squeezed into protons, forming neutrons. If the amount of original matter is on the low end of the range, the process stops here, forming a neutron star—a relatively cold mass of neutrons perhaps ten miles in diameter. (Neutron stars are commonly called pulsars.) If the amount of matter is larger, the collapse does not stop. It continues until all the matter has been compressed to a volume of zero, a point astronomers call a singularity. The result is a black hole.

See STAR; BLACK HOLE; PULSAR; GRAVITY.

SUTTON, WALTER S., 1887–1916, American biologist. See DNA, HOW ITS ROLE WAS DISCOVERED.

SYMBIOSIS—Any close association of two different living species. Although the term is often used only for relationships that are mutually beneficial, this is, strictly speaking, only one of two alternative forms of symbiosis, mutualism and parasitism, in the latter of which only one partner benefits. Parasitism is a one-sided symbiosis, examples of which are legion. Virtually all forms of life live in a symbiotic relationship with bacteria. For instance, human beings rely on bacteria in their intestines to break down some foods into forms that humans can use, and at the same time supply the bacteria with foods of their own. Lichens are a symbiotic association of an alga with a fungus, the cells of one intermingling with those of the other. The alga provides the fungus with nutrients, through photosynthesis, and the fungus provides the alga with protection from drying out and possibly also with minerals. In a broader sense, "symbiotic mutualism" describes the relationship between larger organisms, such as plants that provide food for animals but which depend on the animals for seed dispersal.

SYNAPSE—The gap across which one nerve cell sends its signal to the next nerve cell. The signal is transmitted as a burst of chemicals released by the originating nerve cell. The chemicals are called neurotransmitters.

See NEURON.

SYNERGISM—A fancy way of saying that the whole is greater than the sum of the parts. Synergistic effects sometimes occur with chemical reactions, especially those that happen in the body when drugs are taken. If for example drug A produces 10 units of effect and drug B produces 10 units of effect, the two drugs, if given together, might act separately and produce 20 units of effect. This would not be synergism. Or the drugs might somehow compete and produce less than 20 units of effect, which would also not be synergism. But if the drugs produced more than 20 units of effect, synergism would be at work. This type of synergism, or synergy as it is sometimes called, can occur with all kinds of substances, including toxic chemicals from the environment, so-called recreational drugs including alcohol, and therapeutic drugs.

TATUM, EDWARD L., 1909–1975, American microbiologist. See GENES, HOW THEY WORK.

TAXONOMY—The science of classifying living organisms according to their resemblances and differences. Though Linnaeus devised the taxonomic system used today in the early eighteenth century, it has turned out to be a very close approximation of evolutionary relationships that did not even begin to come clear until well after Darwin's work in the mid-nineteenth century. The basic system, from the largest grouping to the smallest, is as follows: kingdom, phylum (for animals) or division (for plants), class, order, family, genus, species. In actual practice it has turned out that some early classificatory categories were too broad or too narrow and, to correct the situation without upsetting the system, taxonomists have had to create intermediate classificatory categories between these. These go by such names as suborder, superfamily, subspecies, and the like. Human beings, for example, are classified thusly:

Kingdom	Animalia
Phylum	Chordata
Subphylum	Vertebrata
Class	Mammalia
Family	Hominidae
Genus	*Homo*
Species	*sapiens*

Although the original system of taxonomy had just two kingdoms, plants and animals, this was quickly found inadequate because many organisms do not fit either. Today's classification system has five kingdoms, as follows: Monera (bacteria and related forms), Protista (one-celled organisms with nuclei, such as amoebas and paramecia), Fungi (mushrooms, molds, and other fungi), Plantae, and Animalia. In evolutionary terms the monerans were the first to arise. They gave rise to the protistans, from which plant-like forms evolved into true plants, animal-like forms evolved into true animals, and fungus-like forms evolved into true fungi.

See GENUS; LINNAEUS; SPECIES.

TEMPERATURE—A measure of the hotness, not of the heat, of a body. The distinction is subtle but essential. Strictly speaking, heat is a measure of the total amount of kinetic energy (or energy of motion) possessed by all the vibrating atoms and molecules in any given body, whereas temperature measures only the average kinetic energy of the particles. The distinction may be easier to see in an example. A gallon of water at a temperature of 50 degrees has only half as much heat as two gallons of water at the same temperature. In other words, the total amount of energy in the larger volume is greater, although the average amount of energy is the same for both volumes of water.

Temperature is measured in degrees according to one of the conventional scales (Fahrenheit, Celsius, or Kelvin).

Heat, on the other hand, is measured in different units (BTUs, or British thermal units, as well as calories, in the metric system) and is a function of temperature. A BTU, for example, is the amount of heat required to raise the temperature of a pound of water by one degree Fahrenheit. A calorie is the amount of heat needed to raise the temperature of one gram of water by one degree Celsius. A food Calorie (properly capitalized) is really a kilocalorie, equal to 1,000 calories. The calorie content of food is determined by burning it.

See HEAT; THERMODYNAMICS; ENERGY; CONSERVATION OF ENERGY; ENTROPY.

THEORY—This term, like "hypothesis," is subject to many different definitions. In colloquial parlance, the two words are used interchangeably. In a stricter sense, however, a hypothesis is an idea or supposition that is to be tested in an experiment. A theory is a much broader set of ideas that explain a wide class of phenomena. Usually the ideas must survive considerable testing to win the title of theory. However, ardent proponents of a set of ideas will often prematurely arrogate the term to their ideas. A set of related hypotheses that have survived testing may be elevated to the status of theory. Theories that have been abundantly proven are still called theories. Thus we still speak of the theory of evolution or of relativity theory or, for that matter, cell theory or atom theory.

See PART 1, WHAT IS SCIENCE?

THERMODYNAMICS AND HEAT—It is inside us and it surrounds us. It moves atoms and it permeates the universe. It makes volcanos erupt and cars go. It is heat, a phenomenon that has puzzled thinkers at least since ancient Greek times. It was only about a century ago that scientists finally developed a comprehensive understanding of what heat really is.

The biggest question was whether heat was a property of matter or, in fact, some kind of matter in its own right.

The Greeks debated both ideas, suggesting that heat either resulted from the motion of atoms (atom theory began with the Greeks) or that it was a kind of fluid. By the 1700s most scientists accepted the view that heat was a material thing in its own right, a fluid they called caloric.

Heat as a fluid is, after all, plausible. Put a hot object against a cold one and the heat seems to flow gradually from the hotter to the cooler. As the cooler object becomes hotter, it expands, and as the hotter object cools, it shrinks. It is exactly as if a liquid were flowing between the two objects except for two things: You can't see heat, and the weight of the two objects does not change as the mysterious entity flows from one to the other. Physicists of the 1700s dealt with that problem by saying that caloric was not just an ordinary fluid but an "imponderable" fluid, something that could flow but which had no weight.

One of the first challenges to the theory of caloric came toward the end of the 1700s from Benjamin Thompson, an American who was arrested several times before the Revolution as a spy for the English Crown. Thompson emigrated to England in 1775 and began engineering experiments with guns and cannons. As an adviser to the ruler of Bavaria, Thompson became intrigued by the fact that the boring of a brass barrel of a cannon generated tremendous heat, enough to boil water in which the cannon was immersed. If heat were a fluid, where was it flowing from? Both the cannon and the boring tool were cold at the start. Thompson reasoned that heat was actually being created by the action of the tool against the cannon.

Because Thompson's experiments were carefully controlled, he excluded all possible sources of a fluid heat, and advocates of the caloric theory had to abandon this concept. Others would later show that heat is, in fact, a form of energy, and that when material objects are hot it is because their atoms are vibrating more vigorously than when they are cold. The motion actually makes an object expand when heated. In solids the expansion is relatively slight. In gases it can be quite noticeable, since the atoms are freer to move about and literally bounce off one another with greater force as they acquire more energy with heating. The hot gases suddenly created inside a bullet, for example, bounce off the slug part of the bullet with enough force to propel it at a very high speed. When ordinary heated air expands, a volume of it becomes less dense than the surrounding cooler air and floats upward through this "sea" of cooler, denser air. (The phenomenon is known as convection and is one of three ways in which heat energy can move. The others are conduction and radiation.) Thompson understood convection, and his discoveries led immediately to more efficient designs of home heating radiators.

The next advance in understanding heat came in the early 1800s from Sadi Carnot, a Frenchman who studied the steam engines that were then coming into use. A steam engine, like today's gasoline engines, is essentially a machine for using heat energy to create mechanical motion. Although Carnot began by believing in caloric, his analysis led him to suggest that steam engines were devices for converting a kind of motion stored in the fire that heated these engines into the motion of the piston which, like the slug of a bullet, was pushed by the expanding steam. Mere levers, of course, translate the linear motion of the piston into the more useful rotary action of the engine's drive wheel.

Carnot's great insight was in linking heat and mechanical motion (or work, as a physicist would term it). He speculated that the total motive power of the universe was constant and merely changed from one form to another. In this Carnot reached an insight that he could not fully develop before dying at the age of thirty-six from cholera. It fell to a Prussian, Rudolf Clausius, to take the next step, formally articulating a law of nature that has come to be called the first law of thermodynamics— namely, that energy can neither be created nor destroyed, but only changed from one form to another. The advent of Einsteinian physics a century later would modify this law by revealing that under special circumstances energy can be converted into matter and vice versa. Except for atomic reactions, however, Carnot's speculation remains valid.

A more direct understanding of the relationship between heat and work came from a classic experiment performed in 1847 by James Joule (pronounced jewel), an English physicist. He put a paddle wheel in a tank of water, measured the water temperature, and then spun the paddle wheel for a long time. Gradually the water temperature rose. By measuring the rise in temperature and the amount of work done in cranking the paddle wheel, Joule came up with a measure of how much work it takes to produce a given quantity of heat or, as it is called today, the mechanical equivalent of heat. Joule tried various types of work—mechanical, electrical, and magnetic— and found that when he equated the units used to measure each type of work, it took equal amounts of work to produce a given amount of heat.

At about the same time Joule was working, Clausius and William Thomson (later Lord Kelvin), a British mathematician and physicist, independently arrived at one of the most important discoveries of physics, a finding that has come to be called the second law of thermodynamics. This law has been stated in many ways, but they all boil down to expressing one idea: That it is impossible to convert all of a system's heat energy into non-heat forms of energy. Some always gets lost. This is the law that prohibits perpetual-motion machines and is sometimes colloquially stated as, "there's no such thing as a free lunch."

In other words, if you want to use heat to do work, you have to put in more heat energy than you get out in the form of non-heat energy. This was clearly the case in the steam engines that Carnot studied: some of the heat from the steam was converted into the motion of the piston, but a great deal of it went right through the engine and came out in the spent steam, which was still hot. The mathematics of the operation, worked out by Thomson (Lord

Kelvin), showed that it was impossible to convert all the heat of the steam into mechanical energy. This is not true, however, when converting energy in the opposite direction. Non-heat energy, such as mechanical motion, can be fully converted into heat.

The implications of the second law are both commonplace and profound. Consider first a wind-up watch. Gradually it converts all the stored elastic energy of its spring into heat. Initially the elastic energy of the spring turns into kinetic energy when the gears and levers of the watch move, but as they move, the kinetic energy becomes heat in the friction of bearings and gears. The heat radiates into the air and can never be recaptured. When the energy stored in the spring is used up, the watch stops. Even though the watch holds some heat energy, it will never use this heat to wind itself up, not even if the watch is heated. It is necessary to rewind the watch, using the energy of muscle contraction, which comes from the release of chemical energy inside muscle cells. This chemical energy comes from another form of chemical energy in the food one eats, which ultimately comes from the solar energy that plants store during photosynthesis (or from the animal that ate the plants).

The more profound implication of the second law is that without a continuing source of energy from the sun, the energy transformations on the Earth would gradually use up the energy already stored here. All machines and all life would eventually grind to a halt. The Earth would run down. Even the internal heat of the Earth, generated by atomic decay, will gradually dissipate, since this decay cannot go on forever. Eventually, this will happen anyway because the sun will some day exhaust its nuclear fuel and die. So will the entire universe. (It will, that is, unless its expansion stops and reverses, collapsing the universe back to the state it was in before the Big Bang. But nobody knows whether that can happen.)

This ultimate dissipation of energy is called entropy. The second law of thermodynamics holds that in any closed system (a watch without someone to wind it, the

Earth without sunlight, or the universe as a whole), the
total amount of entropy increases. Or, in other words,
the amount of energy available for doing work decreases.
The increase of entropy can be halted or reversed, but
only within limited regions of the universe and only by
importing energy from outside those regions.

On at least one planet, however, there has been a spectacular reversal of the entropic process—the origin and
evolution of life. Sometimes religious creationists cite this
as a violation of the second law of thermodynamics, and
therefore as evidence that life can only have arisen
through a miracle. They overlook the fact that the second
law applies only to closed systems and only to the average
of conditions throughout such systems. The Earth is not
closed; it receives abundant energy from the sun. Under
special circumstances, sunlight can cause atoms to join
into more complex molecules. Molecular evolution, driven
still further by solar energy, has produced living organisms of great complexity—organisms that do mechanical
work using captured solar energy. But every time the organisms do mechanical work, they release the energy they
have stored, ultimately as heat. Life is a only a local exception to the general increase of entropy, but as has been
said, one that cannot last forever. If the Earth and sun
are considered as a closed system, then its average entropy is increasing, for the amount of energy being lost by
the sun far outweighs the local reversals of entropy on
Earth.

See ENTROPY; KELVIN.

THOMPSON, BENJAMIN (Count Rumford), 1753–
1814, American-born British physicist. See KELVIN;
THERMODYNAMICS.

THOMSON, JOSEPH JOHN, 1856–1940, English
physicist—The first subatomic particle to be known was
the electron, and Thomson was its discoverer. He not only
discovered the electron, but determined its electrical
charge, that it weighed only one two-thousandth the

weight of a single hydrogen atom, and that it traveled at 160,000 miles per second when flying out of his experimental devices. Though these achievements sound impressive to modern science-watchers, it should be remembered that Thomson accomplished them in 1897— six years before the Wright brothers invented the airplane.

Thomson's great discovery came during his long career at Cambridge University, most of which was as director of the famed Cavendish Laboratory. Like many physicists, he was intrigued by the strange properties of a device called the Crookes tube. This device, a forerunner of the light bulb, the vacuum tube, and the x-ray tube, was built by William Crookes, an English scientist, mainly so that physicists could study its effects. It consisted of a glass container from which air had been removed but into which electrodes projected. One electrode, the cathode, could be heated, and from it would emanate strange rays that made the glass glow.

Thomson's experiments with the cathode rays established that their path could be deflected by magnets and by an electrostatic field (the same kind of "static" field that builds up in a clothes drier or while one is shuffling across a carpet in low humidity). Experiments with static electricity were well known, and it had long been established that it came in two forms, positive and negative. When Thomson found that positive electrostatic fields (emanating from a charged object) attracted the beam of cathode rays in the Crookes tube, he concluded that they were of opposite charge, since opposite charges attract. Through other such experiments, Thomson established that the rays were in fact a barrage of individual particles. He named them electrons.

In subsequent years Thomson's work would be extended by Ernest Rutherford, a New Zealand-born physicist who succeeded Thomson as head of the Cavendish Laboratory. It would be Rutherford who showed the electron's place in the atom.

Thomson is remembered not only as a scientist but as a

great teacher and leader of other scientists. Scores of prominent scientists made important contributions while working under his direction. Eight of Thomson's students went on to win Nobel Prizes, including Rutherford and Thomson's own son, George Paget Thomson.

See ELECTRON; ATOMIC STRUCTURE; RUTHERFORD.

TIME DILATION—Einstein's relativity theory says that the faster an object moves, the slower time passes for that object as observed by a witness moving at a slower speed. The amount of slowing, or time dilation, is noticeable only when objects move at an appreciable fraction of the speed of light. In no case, however, is this slowing noticeable by a person moving at such a speed unless and until he returns to the slower-moving world and finds that more time has elapsed there than he has experienced.

See RELATIVITY; EINSTEIN.

TISSUE—A large group of cells more or less all of the same kind. An organ, such as a kidney, may be composed of several kinds of tissues. The distinction of a tissue from an organ is not always clear-cut. For example, blood is often spoken of as a tissue, but it is more like an organ in that it is made up of several very different kinds of cells.

See ORGAN.

TRANSCRIPTION—In molecular biology this term refers to the process by which the genetic message encoded in DNA is "transcribed" into a new molecule, made of RNA, which can carry the message out of the cell nucleus and to the cellular machinery, which can in turn read and follow the message.

See RNA; GENES, HOW THEY WORK.

TUNGUSKA EVENT—This is one of the few single-event mysteries in the history of the world that really is a good mystery and really does suggest that something astounding once happened. On the morning of June 30, 1908, a thundering explosion shook a vast region, called Tunguska, of northern Siberia. It blew down trees for twenty-five miles around. The shock waves were recorded all over the world. No people were hurt, but there were charred remains of reindeer in the area and reports of horses being knocked over 400 miles away.

Ever since, scientists all over the world have been trying to figure out what caused the blast. It is estimated to have had a power of at least 10 megatons, the yield of a large hydrogen bomb. Since nobody on Earth had atomic weapons in 1908, there have been theories that an alien culture tried to bomb the earth or that one of its nuclear-powered ships blew up. Less fanciful are the leading theories of today—that an asteroid or a comet exploded just before impact, or that a microscopic black hole penetrated the Earth. The blast had to have occurred before the object hit, because no crater was formed. In fact the only long-term evidence of the blast are the blown-down trees. They all fell away from the center of the blast, and old aerial photos show them lying like so many spokes in a wheel. Trees at the hub of the blast site were not knocked over, but were merely stripped of their branches. Clearly the blast reached them from directly above. The asteroid theory suggests that the Tunguska object weighed around a million tons, and was anywhere from 300 to 600 feet in diameter. Other theories say the object must have weighed more like seven-million tons.

See BLACK HOLE; ASTEROID; COMET.

UNIFIED FIELD THEORY—See GRAND UNIFIED THEORIES.

UNIFORMITARIANISM—Not a religion but a theory of geology asserting that the major features of the Earth, from its mountains to its oceans, have been shaped by gradual processes operating more or less uniformly over very long periods. The theory was developed by Charles Lyell, a nineteenth-century British geologist, as an alternative to the then dominant view called catastrophism. This view, now discredited, claimed that sudden events, such as Noah's flood, had shaped the Earth's surface. Uniformitarianism, by contrast, says that mountains are built slowly, thrusting up bit by bit over hundreds of thousands of years. The Rocky Mountains, for example, are known to be comparatively young and still growing, getting steadily taller. By the same token, erosion gradually wears away old mountains, such as the Appalachians, flattening the landscape and sending billions of tons of silt downstream to create new land at river deltas. The new theory of plate tectonics, which says for example that the Atlantic Ocean is slowly widening, accords nicely with uniformitarianism.

See CATASTROPHISM; LYELL; PLATE TECTONICS.

UNIVERSE—Everything that exists, including all forces, all objects from subatomic quarks to galaxies, and all the space that separates these objects. It is estimated that the universe contains 2.2×10^{41} pounds of matter. This quantity can also be expressed as 220, 000, 000, 000, 000, 000, 000, 000, 000, 000, 000, 000, 000, 000 pounds. Perhaps the most profound modern discovery about the universe is that it has a finite age. A virtually overwhelming body of evidence now indicates that the entire universe was born in an explosion of unimaginable proportions—the Big Bang—between 13-and 20-billion years ago. It is not yet clear how the universe will die.

See BIG BANG; CLOSED UNIVERSE; FORCE; MATTER; ATOM; QUARK; GALAXY; STAR; SOLAR SYSTEM, ORIGIN OF.

UREY HAROLD, 1893–1981, American biochemist. See LIFE, ORIGIN OF.

VALENCE OR VALENCY—Atoms have different powers to form chemical bonds with one another. The power to do so is expressed as a valence number, or valency. It can be thought of as the number of electrons the atom has available to share with other atoms so as to form a stable compound.

See CHEMICAL BONDING.

VESALIUS, ANDREAS, 1514–1564, Belgian anatomist—Vesalius spent much of his youth pursuing the macabre hobby of dissecting dead animals to see how they worked, on the basis of their internal anatomy. Thus he was shocked upon entering medical school to learn that fledgling doctors were being taught human anatomy by way of readings from a 1,300-year-old book based on dissections of monkeys and pigs. The book was by Galen, the famed Greek physician who died in A.D. 200 but whose teachings, though laden with errors, still held sway in a Europe where the Church forbid dissection of human cadavers. Such a method of teaching naturally irritated the young Vesalius, and he resolved to teach himself the inner workings of the human body just as he had taught himself the anatomy of various animals.

One obstacle to this was obtaining bodies to dissect. The records are not totally clear, but it appears that Vesalius either became a body-snatcher himself or hired others to bring him fresh corpses secretly dug up from graves. There were even allegations that Vesalius's ghouls resorted to more direct means of producing fresh cadavers. The end result, however, was Vesalius's seven-volume masterpiece on the structure of the human body, a work of Renaissance scholarship that swept away centuries of myth and superstition and revolutionized medicine, putting the field on an empirical basis, much as Copernicus and Kepler did for astronomy at roughly the same time. Vesalius's accurate descriptions of the various organs and tissues he found in the body immediately raised questions about their function. Vesalius thus made anatomy a science, and revealed the need for a new science, physiology—the study of the functions of internal organs.

Vesalius was born in Brussels to a family of doctors and apothecaries. He studied for his M.D. degree at the universities of Paris and Padua, and upon graduating was immediately given a professorship at Padua. It was there that he obtained his cadavers, performed numerous dissections, and wrote his epochal work, publishing it at the age of twenty-eight.

Vesalius knew his practices would be frowned upon— or worse—by the authorities, and that he might be forced out of the university. So before publishing he sought to assure himself protection from the coming storm. He took a copy of his treatise to the Holy Roman Emperor, Charles V, king of Spain. Vesalius's father had served Charles as court apothecary. The younger Vesalius impressed the emperor and was offered the post of royal physician. Thus protected, Vesalius published his books in 1543 and rode out a furious storm of controversy.

From this point on, Vesalius pursued no more anatomical research, instead devoting himself to the emperor's service for about ten years. He then returned to Brussels, where he operated a prosperous private medical practice until Charles's son, Philip II, now on the throne, called him back to Madrid. Some years later, at the age of fifty, Vesalius made a pilgrimage to the Holy Land, became ill, and died on the return trip.

VIRCHOW, RUDOLF, 1821–1902, German pathologist. See CELL THEORY.

VIRUS—Although sometimes thought of as the smallest living organisms, viruses are not alive, or at least not in the usual sense, for they lack the ability to reproduce by themselves. They consist of two parts: a short segment of nucleic acid (either DNA or RNA) encoding a small amount of genetic information, and a protein coat that covers this genetic material and, depending on the virus, can be shaped as simply as a ball or as elaborately as some fantastic spacecraft. The only way a virus can reproduce is by getting inside a living cell—known as the host cell— and borrowing the cell's protein-synthesizing machinery (mainly its ribosome) to carry out its genetic messages. The messages simply call for the manufacture of thousands of more identical viruses. So many new viruses are made that the host cell's energy and supply of raw materials (amino acids for making proteins) are exhausted. Often, the emerging new viruses destroy the host cell as they move on, each to attack another cell. Some viruses do not kill the host cell, but instead insert their genes into the cell's chromosomes, where they may lurk dormantly for many years before acting. These "slow viruses," as they are sometimes called, may be the agents of many currently mysterious diseases, such as multiple sclerosis.

See DNA; RIBOSOME.

VOLTA, ALESSANDRO, 1745–1827, Italian physicist—The age of electricity began in 1800 when Volta, a physics professor at the University of Pavia, built the world's first battery—the first source of continuous electric current. (Static electricity, which discharges itself instantaneously, had been studied for more than a century.) Although it would be many years before the battery had any practical use, it immediately became a scientific instrument as important as the telescope or the microscope. Volta himself and others used the battery to study both the nature of electricity and to discover new chemical compounds produced through a process called electrolysis.

Volta's work was inspired in 1791 when his friend, Luigi Galvani, a biologist at the University of Bologna, made a surprising discovery while dissecting a frog. As Galvani was touching an exposed leg muscle with a steel scalpel, the upper end of the scalpel happened to bump a brass hook being used to pin down the dead frog's body. As the two metals touched, the leg muscle twitched violently. Galvani repeated the procedure and the muscle jerked every time he did. Galvani published his finding, suggesting that he had discovered a new form of electricity. He called it animal electricity, and much discussion ensued about whether it was somehow part of the mysterious essence of life.

Volta, at first skeptical about the phenomenon, repeated the experiment and became an enthusiastic convert. But not to the idea of animal electricity. As a physicist he was more impressed by the fact that the phenomenon worked only if the two metal objects were made of different metals, and by the fact that certain combinations of different metals produced stronger twitches than others. Zinc and silver produced the strongest response. Along with trying various metals, Volta also made substitutions for the frog's leg. One thing he tried was his own tongue, a structure that is mainly muscle. He placed a strip of tin on the tip of his tongue and then touched the back of his tongue with a silver spoon. Next he brought the spoon's handle down to touch the tin, expecting his tongue to twitch. The contact of the silver and tin didn't so much cause his tongue to twitch as produce a sharply sour taste. This gave Volta the idea that the crucial intermediary between the metals was not muscle but a fluid, perhaps like saliva, capable of conducting the electricity from one metal to the other.

After more experiments, Volta hit upon brine, a strong saltwater, as the most effective conducting fluid. A piece of cardboard soaked in brine nicely substituted for the frog's leg or saliva. Volta even found that the amount of electricity produced could be increased by piling up a stack of dissimilar metal disks and sandwiching a brine-soaked piece of cardboard between each. The device, eventually called a Voltaic pile, was the world's first battery (each sandwich being one cell) and it could make muscles twitch far more strongly.

Electricity, Volta proved, was not an inherent property of animal tissue, but instead was a result of the contact of dissimilar metals. Twitching was an effect, not a cause of electricity. (See CELL [ELECTRICAL] for an explanation.) Having settled his debate with Galvani, Volta published his invention and findings in a letter to London's Royal Society, then the most prestigious scientific body in the world. It was an immediate sensation, and Napoleon invited Volta to Paris for a demonstration of the battery. The emperor was so impressed that he awarded Volta the medal of the Legion of Honor and made him a count and a senator in the Kingdom of Lombardy, Volta's home territory, which Napoleon had just conquered. A different kind of honor would come in 1893, sixty-six years after Volta's death, when the international Congress of Electricians named the unit of force that drives an electrical current (measured in amperes) the volt.

Volta was born in Como, a beautiful village at the foot of the Italian Alps. His family was highly religious, almost all of his male relatives becoming priests. Volta, however, preferred science, and especially electricity, which was then known only in the form of static electricity. He earned his university degree at the age of seventeen and got a job teaching physics at the Como high school. At the age of thirty-two he was asked by the University of Pavia to establish a physics department there. It was at Pavia that Volta invented the battery, an achievement that also prompted the emperor of Austria to give Volta the job of director of the philosophical faculty at the University of Padua. Volta accepted, and in 1819, at the age of seventy-four, retired and returned to Como. He died there at the age of eighty-two.

See CELL (ELECTRICAL).

WALLACE, ALFRED RUSSEL, 1823–1913, English evolutionary biologist. See DARWIN.

WATSON, JAMES D., 1928–   , American molecular biologist—It was a toss of the coin in 1953 that allowed Watson to put his name first on a paper he had authored with Francis Crick, the English biophysicist. The two had deciphered the double-helix structure of the DNA molecule and, from that moment on, the names of Watson and Crick would be linked in scientific history.

Watson was born in Chicago, and at Indiana University did early research on viruses that confirmed the role in heredity of DNA—then known only as one of many chemicals in cells. In Copenhagen he studied biochemistry and learned of work at England's Cavendish Laboratory involving the use of x-ray pictures of crystals to learn the internal structures of the molecules making up the crystals. Thinking that this method might shed light on the structure of DNA, Watson got a job at the Cavendish Laboratory, where he teamed up with Crick, who already had an interest in determining the three-dimensional structure of molecules.

Following their discovery that DNA is shaped like a double helix—two entwined corkscrews—Watson returned to professorships in the United States, where he worked on efforts to decipher the genetic code carried by the DNA molecule. He wrote *The Double Helix*, a controversial, highly personal account of the epochal discovery of DNA, and which revealed the pursuit of science to be less than strictly high-minded. In 1968 Watson became director of the Cold Spring Harbor Laboratory on New York's Long Island. It is today a major center of molecular biology research.

See DNA; DOUBLE HELIX; GENES, HOW THEY WORK; CRICK.

WAVE—Some waves are fairly easy to understand, while others are among the deepest mysteries of science, lying at the heart of the effort to understand the nature of physical existence itself.

The easy waves to understand are the classical waves, such as water waves and sound waves. Both can be understood as disturbances in a medium. In both cases, the only thing that moves is the disturbance itself. When a water wave passes, water molecules go up and down. The thing that moves sideways over great distances in the water is the *region* in which the water molecules are moving up or down. This comprises an ever-changing population of water molecules as the wave moves from one area to another. Water waves are confined to the two-dimensional surface of the water body, and so can spread only as ever enlarging circles. Sound waves, by comparison, spread in three dimensions as ever enlarging spheres. In a sound wave, the air molecules are momentarily compressed and decompressed. The air molecules do not move more than a fraction of an inch; all they do is absorb an impact from a neighboring molecule and, briefly accelerated by the energy from this blow, slam into the next air molecule, transferring the energy once again. What moves is the boundary of the enlarging sphere, a region where the air pressure is higher because the air molecules are closer together (during the brief interval when one is hitting another).

The waves that are hard to understand are those of light and the other kinds of electromagnetism (such as radio waves), and the wave-like behavior of various subatomic particles. The source of the difficulty is that this wave-like behavior occurs with phenomena that seem not to be disturbances in a medium, but particles instead. Light, to cite a classic example, can be thought of as both a stream or flood of particles (called photons) and as a disturbance in a medium (except that there is no detectable such medium, or ether, as was once supposed). Depending on what kind of experiment one does with light, the results make sense only if light is made of particles, or only if light is a disturbance in a medium, and involves no traveling particles. All electromagnetic waves are like this, and physicists have had to accept that electromagnetism exhibits an inexplicable wave-particle duality.

See PARTICLE; LIGHT, THE NATURE OF; PHOTONS; MICHELSON-MORLEY EXPERIMENT; RELATIVITY.

WEAK FORCE OR INTERACTION—One of the four fundamental forces in nature. It operates only within atomic nuclei that are undergoing radioactive decay, and otherwise plays no significant role. The other three fundamental forces are the strong force, which binds neutrons and protons into the nucleus; electromagnetism, which binds electrons to the nucleus; and gravity.

See FORCE; UNIFIED FIELD THEORY.

WEGENER, ALFRED, 1880–1930, German geologist— Alfred Wegener is the author of the modern theory of continental drift—or plate tectonics, as it is also called— but when he produced his theory in 1915, it was soon rejected by his colleagues, and Wegener died never knowing that it would eventually be accepted as one of the greatest advances in geology: a theory, now backed with much evidence, that explains most of the major geological features of the Earth's surface. Wegener disappeared while on a geological expedition to Greenland, seeking more evidence for his theory, and was never heard from again.

Wegener was born in Berlin and trained as an astronomer. He also taught meteorology and geophysics at the University of Graz in Austria.

For a full explanation of his theory, see the entry on PLATE TECTONICS.

WEIGHT—The phenomenon that results when gravity acts on mass.

See MASS; WEIGHTLESSNESS.

WEIGHTLESSNESS—It has been more than a quarter of a century since the Soviet Union lofted the world's first artificial satellite into orbit. Ever since that fateful day in 1957, millions of people have been content to believe that Sputnik and its many companions in space stay up because they have somehow freed themselves from the effects of gravity.

The fact is that Earth's gravity, and that of all other celestial bodies, extends out infinitely. You can never completely escape it no matter how high you go. It gets weaker the farther you go from the planet, but it is not appreciably diminished at the altitude traveled by artificial satellites. Earth's gravity, it should be obvious, is what keeps the moon in orbit.

Satellites stay up precisely because of the effect of gravity, combined with their forward speed. Actually, satellites are always falling, pulled downward by gravity. They never hit the ground because, at the same time, they are coasting forward in a direction tangent to the Earth's circumference. The two motions—down and forward—combine to produce a curved path that is essentially parallel to the Earth's curvature.

Consider what happens when you throw a ball forward. An observer standing to one side can see that as the ball moves forward, it is also dropping toward the ground. It follows a curved path. The harder (faster) you throw the ball, the shallower the curve. If you could throw the ball fast enough, the curve would be so shallow that it would parallel the curvature of the Earth. The ball would be in orbit. Of course, wind resistance would soon slow the ball down until it lost the forward speed needed for it to stay in orbit. Above the atmosphere, this friction is gone, and there is nothing to slow a satellite's forward motion.

Astronauts inside an orbiting spacecraft are weightless only in the sense that they are moving at the same speed, both forward and down, as the spacecraft and everything in it. Skydivers are equally weightless. They just happen to be moving in a path that will eventually bring them to the ground. A weightless astronaut still has mass. Mass is what will raise a bump on his head if he drifts until he hits the spacecraft wall. It is also what will kill the skydiver if his parachute does not open.

See MASS.

WEINBERG, STEVEN, 1933– , American physicist. See GRAND UNIFIED THEORIES.

WILSON, ROBERT W., 1936– , American astrophysicist. See BIG BANG.

The Author

Boyce Rensberger has been preparing science information for non-scientific readers for almost twenty years. He began in 1966 as science writer for *The Detroit Free Press* and moved to *The New York Times* in 1971. After eight years at *The Times*, he left to freelance science articles to more than a dozen national magazines, and to help create "3-2-1 Contact," an Emmy Award-winning science television series for children. For three years Rensberger was senior editor of *Science 81/84*, which won the National Magazine Award for general excellence in two of those years. In 1984 Rensberger joined *The Washington Post* to resume newspaper science writing. His newspaper articles over the years have won many national and local awards. One award, the Alicia Patterson Foundation Fellowship, sent him to Africa for a year to cover research on wildlife conservation and human evolution. That year led to Rensberger's previous book, *The Cult of the Wild*.